THE MOTION PARADOX

THE
MOTION PARADOX

The 2,500-Year-Old Puzzle
Behind All the Mysteries
of Time and Space

JOSEPH MAZUR

DUTTON

DUTTON
Published by Penguin Group (USA) Inc.
375 Hudson Street, New York, New York 10014, U.S.A.
Penguin Group (Canada), 90 Eglinton Avenue East, Suite 700, Toronto, Ontario M4P 2Y3,
Canada (a division of Pearson Penguin Canada Inc.); Penguin Books Ltd, 80 Strand,
London WC2R 0RL, England; Penguin Ireland, 25 St Stephen's Green, Dublin 2, Ireland
(a division of Penguin Books Ltd); Penguin Group (Australia), 250 Camberwell Road, Cam-
berwell, Victoria 3124, Australia (a division of Pearson Australia Group Pty Ltd); Penguin
Books India Pvt Ltd, 11 Community Centre, Panchsheel Park, New Delhi – 110 017, India;
Penguin Group (NZ), 67 Apollo Drive, Mairangi Bay, Auckland 1311, New Zealand (a divi-
sion of Pearson New Zealand Ltd); Penguin Books (South Africa) (Pty) Ltd, 24 Sturdee
Avenue, Rosebank, Johannesburg 2196, South Africa

Penguin Books Ltd, Registered Offices: 80 Strand, London WC2R 0RL, England

Published by Dutton, a member of Penguin Group (USA) Inc.

First printing, April 2007
10   9   8   7   6   5   4   3   2   1

 REGISTERED TRADEMARK — MARCA REGISTRADA

LIBRARY OF CONGRESS CATALOGING-IN-PUBLICATION DATA HAS BEEN APPLIED FOR.

ISBN 978-0-525-94992-3

Printed in the United States of America
Set in Granjon
Designed by Victoria Hartman

To Jennifer
In memory of her mother,
my dearly beloved mother-in-law,
Anne Joffe—
perceptive questioner, astute listener,
kindhearted caregiver.

Contents

Part 4 · A Commotion of Absurdities, Revisited

THE MOTION PARADOX

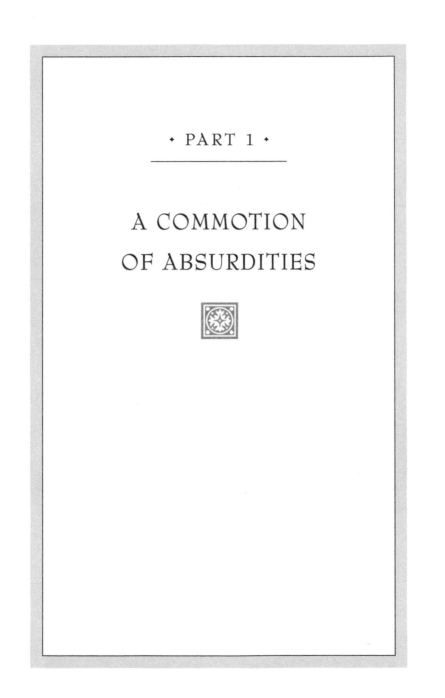

· PART 1 ·

A COMMOTION
OF ABSURDITIES

Preamble to the Paradoxes of Motion

My father was the first person to tell me about paradoxes of time. He had never heard of Zeno's paradoxes, those peculiar arguments on motion that contradict common sense and that have been misunderstood these last two and a half millennia, but was a gentleman philosopher with instinctive wisdom about the world and how it turned. My brother had just received a brand-new Schwinn bicycle with chrome fenders, a speedometer, and battery-operated horn for his birthday. Boy, was that neat. The gentleman philosopher knew just what I was thinking. To soothe my jealousy, he took me aside and told me that I was half my brother's age, but in eight years I would be three-quarters his age and that from then on there would hardly be a difference. Of course, I had no idea what he meant by three-quarters, let alone three-quarters of someone's age. When I asked how old would I have to be to catch up completely, he laughed and said that that would never happen, but that the difference would always be getting smaller. Years later, I thought I understood; but, now, rapidly gaining on my brother as I pass sixteen-seventeenths of his age, I'm just beginning to. Incidentally, my brother's bicycle was stolen shortly before his next birthday.

More than 2,000 years before my father eased my bike envy with his thought experiment, Zeno had invented similar paradoxes. Zeno argued with flawless logic that, contrary to what everyone experiences every day, nothing moves.

Zeno's four paradoxes listed in Aristotle's *Physics* are:

> The Dichotomy—*That a moving object will never reach any given point, because however near it may be, it must always first accomplish a halfway stage, and then the halfway stage of what is left and so on, and this series has no end. Therefore, the object can never reach the end of any given distance.*
>
> The Achilles—*That the swiftest racer can never overtake the slowest, if the slowest is given any start at all; because the slowest will have passed beyond his starting-point when the swiftest reaches it, and beyond the point he has then reached when the swiftest reaches it and so on....*
>
> The Flying Arrow—*That it is impossible for a thing to be moving during a period of time, because it is impossible for it to be moving at an indivisible instant.*
>
> The Stadium—*That half a given period of time is equal to the whole of it; because equal motions must occupy equal times, and yet the time occupied in passing the same number of equal objects varies according as the objects are moving or stationary. The fallacy lies in the assumption that a moving body passes moving and stationary objects with equal velocity.*

The flying-arrow paradox concludes that motion is impossible. Zeno pictures an arrow in flight and considers it frozen at a single point in time. He argues that the arrow must be stationary at that instant, and that if it is stationary at that instant then it is stationary at any—and every—instant. Therefore, it does not move at all. This single paradox may bewilder, but the four together release a commotion of absurdities, profoundly questioning our models of reality.

Zeno's paradoxes raise a fundamental question about the universe: Are time and space continuous like an unbroken line, or do they come in discrete units, like a string of beads? It's a question that even today's physicists, who are reputed to be closer than ever to a theory of everything, are struggling with.

Zeno's arguments seem absurd. We know the arrow flies through the air, yet we may have some difficulty in explaining why or how we know. One may argue that the whole notion of fixing a point in time is absurd and that it makes no sense to say that an arrow appears stationary at any point in time. In mathematics, time is a variable that can be fixed by simply declaring it to be some number. We have formulas that tell us where the arrow is at any time t, so if we let t equal some specific time, then we should know the exact spot where the arrow is at that time. Yet this means that our mathematical models of motion, space, and time are merely intellectual constructions built for the convenience of easy calculations, not for the greater purpose of representing the structure of reality.

As we came to understand motion through math with greater sophistication, we shed light on Zeno's paradoxes. But only by solving the ultimate mysteries of time and space can we definitively solve the puzzles that Zeno put forth at the very dawn of science. He was ahead of his time.

HISTORY WAS NOT always generous to Zeno's inventions. At times during the past 2,000 years, his paradoxes were considered nothing more than picky sophisms of logic with little merit for continued discussion. At other times they were considered embarrassments to mathematicians' investigations of infinity and the continuum; our historians tell us that those paradoxes contributed to the Greek abandonment of such investigations.

Almost all of what we know about Zeno's life is speculation, composed from fragments and historical sources written almost a thousand years after his death. We know that he wrote a magnificent book on philosophy that was used as a textbook at Plato's Academy, but not even the smallest fragment of it has survived. The fifth-century philosopher and mathematician Proclus, our principal source of information about the early history of Greek geometry, tells us that Zeno wrote a book containing forty paradoxes, but that it was stolen before it could be published. The four known paradoxes come to us by way of Aristotle alone. Dozens of major works written by renowned scholars from Plato to Bertrand Russell have pondered the paradoxes. This literature contains a plethora of magnificently arching connections across history.

The absence of Zeno's writings warrants suspicion over whether or not the man actually existed beyond merely being a character in Plato's *Parmenides*. Despite that absence, a great deal of extant material tells of his profound philosophical ideas, and one can gather enough from them to assemble a coherent story. Plato and Diogenes Laertius provide the corners to the jigsaw puzzle of Zeno's life, Aristotle and Proclus give the edges of his philosophy, and then we fill in the rest with supposition.

After the death of Archimedes in 212 BCE, the topic of motion was effectively abandoned; it did not resurface for another 1,400 years, when Gerard of Brussels revived the mathematical works of Euclid and Archimedes and came very close to defining speed as a ratio of distance to time. A hundred years later, four Merton College mathematicians sharing ideas on the mechanics of motion were able to work out the first formulas linking acceleration to distance for a freely falling object. It has been claimed that the same math used by the Merton mathematicians solves the Achilles paradox. I'll show that while this may seem to be the case on the surface, the math in question—basic algebra—does nothing to address the underlying phenomenological problem that the paradox drives at.

Three hundred years after the Merton mathematicians, Galileo began to experiment with physical objects to measure their movement, initiating a shift toward an empirical approach to science that is still with us today. It is through Galileo that the connection between math and the physical world became solidified. Newton, Leibniz, and other mathematicians took this approach further and invented the mathematical field we now know as calculus in order to model motion.

Newton had the inspired idea that acceleration, the rate of change in velocity, was completely determined by two entities that have no apparent connection to motion—force and mass. It seemed to many that, at last, motion had been fully explained.

Math had triumphed in the explanation of the physical world. It seemed that calculus could explain the dichotomy paradox. But again, the math is merely a tool. The underlying reality that the paradox addresses is evaded.

Before the eighteenth century, time was crudely measured. Galileo used his own pulse as a measure. Today, our atomic clocks can measure a time interval as small as one-millionth of a second. (Though we have a word for one-billionth of a second— *nanosecond*—we still have no way of accurately measuring it.) But no matter how finely calibrated our clocks are, they are always measuring something discrete—an interval, a repeating signal, a duration between events. This is the heart of the problem: We measure time as a duration and think of motion as continuous. The best definition of motion we have is intricately tangled between the discrete and continuous impressions of time and space. Despite contributions by Aristotle, Galileo, Newton, and many others, for over 2,000 years nobody offered better clues about motion's deeper nature than Zeno.

The twentieth century brought relativity and quantum mechanics. Space and time were no longer thought of as separate aspects of reality; they were united into a single four-dimensional continuum. Time dilation, inconstancy of mass, and special relativity suggest that motion is indeed illusory. Motion changes mass—or is it the other way around? Quantum theory suggests that some motion is not continuous. Electrons cannot just sit anywhere within an atom. They are strictly confined to moving between discrete energy levels around an atom's nucleus. Yet we still have a hard time imagining them discretely jumping around, disrespecting our sense of continuous motion. One can't help imagining Zeno rejoicing as his paradoxes return, no longer cast off as answered by simple calculus arguments.

One thing is sure: Everything in this universe, every atom, every molecule, is in some form of motion, whether it be simple locomotive displacement from one place to another, random molecular bombardments, or complex, astonishingly fast, unavoidable vibrations of energy transfer. And our understanding of that motion remains fundamentally paradoxical. How we have pursued the mystery of motion, and all the technological and scientific advances that pursuit has enabled, is one of the greatest stories of our civilization.

· 2 ·

Zeno's Visit to Athens

 Athena was the gray-eyed goddess of war, fertility, art, and wisdom. Her birthday was one of those rare days when women and freed slaves were permitted to appear leisurely in public places. Imagine being in sight of the majestic Acropolis near the northwest corner of the great Athens market and gathering place. Looking southeast along the Panathenaic Way, the dusty path partly shaded by poplars and wild, hardy carob trees, you would see preparations for the Great Panathenaea festival. You would see athletes rubbed with olive oil competing for prizes in foot races, boxing, long jump, javelin throwing, and chariot racing; musicians competing with voice, kithara, and flute; and blind bards reciting Homer's epics. On this day in 450 BCE it was four years since the last great festival, just one year after the signing of a five-year truce between Athens and the other regional power, the city-state of Sparta.

Northwest, past the marketplace, through the sacred gate of the city wall and to the right lay the Ceramicus, a public square and war cemetery in the potter's district. Pentelic marble stones were being stored for the anticipated construction of the Theseion, a temple to honor Hephaestus, the skilled fire-god of the anvil with huge bulk, thin legs, "sturdy neck and hairy chest." It was quieter there, away from the loud hawking butchers, bakers, apiarists, olive pressers, wine merchants, and ironmongers lining the crushed limestone avenue leading to the festival high on the hill. Wild thyme grew through the limestone cracks near fruit vendors selling pears and figs. As Homer noted in *The Odyssey*, there, such fruit "comes at all seasons of the year and there is never a time when the West Winds' breath is not assisting, here the bud, and here the ripening fruit: so that pear after pear, apple after apple, cluster on cluster of grapes, and fig upon fig are always coming to perfection."

According to Plato, Antiphon the Sophist heard the story of Zeno's visit to Athens from his friend Pythodorus so many times that he could repeat it by heart. Parmenides, founder of the celebrated Eleatic school of philosophy, was sitting on a stone, a distinguished man in his sixties with bone-white hair. Sitting next to him was Pythodorus, a younger bearded philosopher looking particularly alert. Next to him was Aristoteles, a sunbronzed man in his thirties, lost in contemplation, and young Socrates, not yet twenty. A nearby donkey was obstinately complaining about a load of barley on its back.

Zeno of Elea, a "tall and attractive" intellectual revolutionary, was reading from his famous book on philosophy. He had come to Athens from Crotona in southern Italy with his teacher and lover Parmenides to visit Pythadorus in the Ceramicus just outside the city wall and to attend the great festival. His lines of reasoning were terribly confusing; they seemed to rely on language tricks aimed toward the mystifying suggestion that there is only one single thing in this world—the thing he called Being—and that all else is mere appearance. He argued that if a thing can be divided, its divided parts can also be divided and such divisions can continue indefinitely. From this he concluded that change, and hence motion, is not possible. He finished reading, but his audience was confused. Even Socrates was confused. He called out to Zeno.

"Zeno, what do you mean? 'If things are many,' you say 'they must be both like and unlike. But that is impossible; unlike things cannot be like, nor like things unlike.' That is what you say, isn't it?"

"Yes," replied Zeno.

The rest of his audience was as bemused as Socrates, who said, ". . . your exposition . . . seem[s] to be rather over the heads of outsiders like ourselves." Zeno was suggesting connections between the problem of plurality, being, continuity, and motion.

We have heard it all before. "And God made the firmament, and divided the waters which were under the firmament from the waters which were above the firmament: and it was so. And God called the firmament Heaven." In the book of Genesis, from the waters came two distinct things—heaven and earth. Creation is division to mark opposites—light and darkness, day and night, summer and winter, land and sea, fish and fowl, even and odd, good and evil.

What Zeno said makes sense. If two things exist, a third must exist to separate them, otherwise there would not be two things, only one. If three things exist, a fourth and fifth must exist to separate the three. To distinguish between A and B there must be a separator C, and to distinguish between A and C there must be another separator D and so on, thus proving that there must be either only one thing in this world or an infinite collection of things. "So," Socrates continued, "are you giving just one more proof that two things do not exist? Is that what you mean, or am I understanding you wrongly?"

"No," answered Zeno, "you have quite rightly understood the purpose of the whole treatise."

Zeno went on to argue that nothing changes because change would require a *becoming* and an *end to being.* "Therefore," Parmenides said, "the one which is not, not possessing being in any sense, neither ceases to be nor comes to be." He and Zeno were thinking that something in an act of change must perform that act *in time.* So change is equivalent to motion; like the arrow that can never leave the bow, change is impossible.

Zeno's arguments for motion may also be applied to the ripening of a pear. The neurologist Oliver Sacks once wrote, "I would come down to the garden in the morning and find the hollyhocks a little higher, the roses more entwined around their trellis, but, however patient I was, I could never catch them moving." We have all seen a garden of flowers, but have we ever seen the flowers growing? Like the hollyhocks, we can never catch a pear ripening, and though it may change in color, taste, texture, and even shape, it remains a pear. How does the pear get from unripe to ripe if every instant we look at it, it is in a fixed state somewhere between two extremes? Zeno's paradoxes are not only about locomotion but also more generally about change in quality and quantity.

· · ·

ZENO WAS A citizen of Elea, a poor Greek colony in what is now southern Italy, when Greek colonies were spreading in all directions to the banks of the Mediterranean like driftwood. Elea was "possessed of no other importance than the knowledge of how to raise virtuous citizens." Long before Alexander the Great conquered regions as far west as Marseille and as far east as India, Greece had established colonies from Carthage in North Africa to Nazareth in Palestine. In a few active centuries a small number of Greeks had developed an enormous intellectual culture connecting politics, the arts, and philosophy. They created a system of government in which a state's affairs were not simply the private interests of the king or governor, but the collective interests of its people, an experiment in democracy. Music, politics, and art combined to inspire Sophocles, Aeschylus, and Euripides to write plays of humor, tragedy, and philosophy for crowds as large as 17,000 in the Athenian outdoor theater. The Greeks discovered the mysteries of number's nature, which led them to the beginnings of what we, today, call mathematics.

Pythagoras of Samos, who lived from about 560 to 480 BCE, was probably the most famous and charismatic mathematician of the time. We know very little about him, but that he traveled widely in the Greek world and settled in Crotona on the southeastern end of the Italian peninsula. His mathematics had a mystical aspect that drew a group of devoted students, a sect of disciples, a brotherhood that lasted for a century after his death. The Pythagoreans influenced many, including Zeno. In particular, the notion that lines were made from strings of points like threads of miniscule beads beguiled him. However, Zeno and Parmenides refuted that Pythagorean notion, and argued that if a line were made of a finite number of points, then time, too, must be built from a finite number of instants and the days would pass not in a smooth continuous flow but in discrete increments, each like a grain of sand falling in an hourglass. This was a time when growing educated classes were strongly aware of Pythagorean discoveries and their ramifications for science and geometry.

The Pythagorean brotherhood's discovery of the connection between the sizes of the sides of a right triangle blurred number theory's bond with geometry and, at the same time, gave one of the first inconsistencies of a mathematical modeling of the physical world. The Pythagorean theorem states that the sum of the squares of the lengths of the sides of a right triangle equals the square of the length of the hypotenuse. This beautiful little theorem eventually caused enormous philosophical problems for the Pythagorean brotherhood, which believed that number represented all things in this world. Legend has it that the Pythagoreans sacrificed an ox on their discovery of the famous theorem (though it's hardly likely that a strictly vegetarian cult with a belief in soul transmigration would do so).

Pythagoreans believed that everything in the world could be represented by finite arrangements of whole numbers. The number 2 represented opinion, 3 signified harmony, and 4 stood for justice. Odd numbers were male, even numbers female. So the number 5 symbolized marriage, because it was the sum of the first even number with the first odd number. The number 10 was holy because it was the sum of the generators of special dimension, $1 + 2 + 3 + 4 = 10$. The number 1 establishes a reference point, 2 points determine a unique line, 3 points not on a line determine a unique plane, and 4 points determine a tetrahedron in space. All numbers were either whole (1, 2, 3, etc.) or rational (fractions of whole numbers).

We are probably missing a lot about Pythagoras, since a covenant bound the Pythagoreans to secrecy over their master's teachings and anything else taught or discovered by the brotherhood, and moreover, the history of Greek civilization before Plato's time is murky. One of their secrets was the construction of the regular pentagram, the five-pointed star and symbol of the brotherhood that comes from connecting the corners of a pentagon. This *cosmic figure*, as the Greek historian Proclus later called it, is not easy to construct if the only tools permitted are a straight edge and compass, or, in other words, straight lines and circles. An isosceles triangle, with one angle equal to four-thirds one of the others, must be constructed. Such a triangle would have an angle of 72 degrees, and that is exactly what is needed to complete the pentagon (because a pentagon has five sides and the sum of all the angles of construction of the regular pentagram is 360 degrees).

Imagine the power these people felt upon discovering how to construct the pentagon, the five-sided figure that leads to an infinite nest of shrinking replicas of itself and an infinite expansion of growing replicas, along with its powerful numeric and geometric qualities.

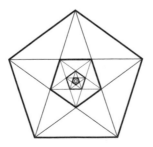

The ratio of a side to a diagonal of such a pentagon gives rise to the golden mean, a number that continues to have spiritual significance among aficionados attempting to discover its hold over nature. These were also folks who believed that gods in human form watched over the actions of individuals, families, and states. From the beginning of the sixteenth century, the golden mean, whose name was not to be coined until the nineteenth century, has been considered a divine proportion because of its ubiquitous presence in the natural world and also because of how it connects simple finite constructions with infinity.

Numerical patterns also suggested to Pythagoreans that numbers were the clues to understanding the nature of the physical world. They saw numbers in music when they discovered that a plucked string produces the same note (one octave higher) as a string twice its length, and extended music theory to a harmony of the soul. They saw numbers in nature, observing the fine structures of flowers. They saw numbers in the construction of their temples, where form followed what they considered to be the spiritual beauty of divine number relationships. They saw numbers in sculpture and art as their artists sought to represent the general makeup of shared attributes, rather than the soul of an individual. They saw numbers in their plays, built on structured themes of crimes and curses. All this logic, structure, and clarity, all this love of symmetry, form, and perfection was applied to reasoning and a belief that the universe is ordered and explainable.

Math was in its youth. The invention of negative numbers would have to wait almost another 800 years for Diophantus to first mention them in his book *Arithmetica* after he found $x = -4$

to be the *absurd* solution to the equation $4x + 20 = 4$. Such absurd solutions would have to wait another 500 years before the Indian mathematician, Mahavira, actually used them and gave them a noble place in number theory. Zero had not been discovered, and neither had tomatoes, tobacco, or coffee (wine was the drink of choice, though goat's milk was tolerated).

The discovery of the Pythagorean theorem inevitably led to the discovery of incommensurables. What if you have a square with sides of length 1? The size of the diagonal would be the square root of 2. But the square root of 2 cannot be written as a ratio of two whole numbers. It is not 7/5, nor 10/7, although they are rough approximations of the square root of 2. No whole number can be divided by another to give the square root of 2. For people who worshiped number, this was extremely unnerving. Anyone discovering relationships such as $1^2 + 3 = 2^2$, $2^2 + 5 = 3^2$, $3^2 + 7 = 4^2$, etc., might conceive mystical notions of the powers of pattern and credit them to some deity's impressive wisdom of order. Essentially, one ruler cannot measure both the side and the diagonal of a square. These early Greeks had discovered an immeasurable part of space. Zeno surely knew about this discovery when he posed his paradoxes questioning the continuity of space and time.

Later, in the early part of the twentieth century, Bertrand Russell wrote, "The problem first raised by the discovery of incommensurables proved, as time went on, to be one of the most severe and at the same time most far-reaching problems that have confronted the human intellect in its endeavor to understand the world."

• • •

PHILOSOPHIZING GREEKS OF the fifth century BCE continued a 200-year attempt that began with Thales of Miletus to articulate a more scientific system of knowledge, to reject any supernatural explanations of nature, and to question the essence of things. Rational criticism and debate replaced speculative thought and established myth. Thales believed that the earth rocking on water caused thunder. His attempt to explain the nature of thunder might be called primitive because it rested on false hypotheses, but modern because it dodged the popular belief in the supernatural.

A theory of the atom, albeit crude, was suggested by the Pythagoreans and developed by Anaxagoras, author of a book reputed to be a complete account of the natural world (now, sadly, lost). The argument that complex things must be made of simpler things was further advanced by Empedocles, a rich doctor from the island that is present-day Sicily. He saw those irreducible things as earth, fire, air, and water, but was careful to point out that each of these elements stood for a wide variety of substances. Water, for example, was a term applied to liquids such as molten metals as well as drinkable fluids. Air would have meant any gas, including those expelled from cattle in the fields. All this makes almost modern sense if one views the classification of matter as solid, liquid, gas, and heat.

Heat? Is that matter? Fire seems to be more of an action. Fire can be used to change the three states of matter or combinations of them into the things we see, or to change one state to another—ice to water, water to steam. Empedocles, in his wisdom, listed three material things together with a device for combining, shaping, and altering those material things. Without fire, the world of things must rely on accidental collisions and linkages to change. With it, the local smithy can learn the art of Hephaestus to hammer the world into new shapes and things from the elements. Empedocles says it this way:

> Just as painters, when they decorate offerings—
> men well taught by skill in their art—
> take the many-colored pigments in their hands,
> and, harmoniously mixing them, some more some less,
> make from them shapes resembling all things,
> creating trees and men and women
> and beasts and birds and fish that live in the sea
> and even gods, long-lived and highest in honor:
> so let not deceit persuade your mind that there is any
>   other source
> for the countless mortal things we see.
> But know this clearly, having heard the tale from a god.

Again, that begs the question of what we get when we take a very close look at the elements and see them, even if we have to rely on imagination, as indivisible things of incredibly small size. Atomism holds that all things consist of substances so small they escape our senses. These indivisible *atoms* are thought to be of many forms, shapes, and sizes, becoming perceptible only after massive collections of them entangle, hook, and bind together through motion and collisions in the void. What they become depends on their shape, arrangement, and position. These groupings of atoms can make the imperceptible perceptible, but they can also untangle and unhook to make the visible invisible. There is an astounding resemblance between this atomic theory and our own twenty-first-century one, where we believe that all matter is composed of atoms and that we only see the matter when enough atoms are compounded to make a substance visible. We see gold when there are enough gold atoms to make the collection of gold atoms visible.

Fifth-century BCE atomic science was imaginative opinion, supported by dialogues against equally creative alternative theories. There were no measurements of atomic weight, nor were there instruments to examine matter any finer than what could be seen with the best pair of eyes, but there were consequences that led to further questioning.

Leucippus, the fifth-century BCE Greek philosopher whose thinking was very much influenced by Zeno and Parmenides, was the founder of the first atomic theory of matter, asserting that atoms consist of imperceptibly minute and indivisible particles that differ only in shape and position. This wonderful theory, which was developed later by his pupil Democritus and led to unexpected results in science, bears directly on the Pythagorean trouble with measuring the diagonal of a square. It is likely that the Pythagoreans thought of a line as a string of atoms, so a line twice as long would contain twice as many atoms. Given that, there must be a definite ratio between any two lengths, because the number of atoms on each line must be finite and hence the ratio of lengths must be a fraction whose numerator is the number of atoms contained in one line and whose denominator is the number of atoms contained in the other.

The atomist argument is that there is a difference between the physical atom and the geometrical point. The atom is indivisible and indestructible, whereas the point is an imagined notion with no physical substance. They reasoned that material substances could be divided as finely as humanly possible, and from there, imagined a moment when no further division would be possible. "Take a wooden stick," they reasoned. "Cut it in two parts. And every day cut the longer piece in half. Continue this cutting day by day, indefinitely. One day it will be hard to claim that the longer end is still a stick, yet easy to maintain that it is still a piece of wood. But how many days will pass before the wood becomes non-wood?" Even the smallest speck of sawdust is still wood.

Anaxagoras knew the group of men who gathered in the Ceramicus in Athens to listen to Zeno, and he was a good friend of Euripides and Pericles. He wrote a book on physics, his only book, offering a complete account of the natural world, arguing that there is a bit of everything in everything. How does a human hair grow from nothing? The answer, Anaxagoras would suggest, is that the food digested by the human already contains hair and everything else within it, imperceptible to our senses. According to Anaxagoras, wood, even in the form of minute particles of sawdust, contains a bit of every other substance, including human hair, a notion stemming from the philosophy of Empedocles and Heraclitus of Ephesus declaring that materials might be changed but not destroyed. These men might have wondered how oil disappears from an oil lamp burning through the night, but, had they imagined the answer, they would have foretold our modern conservation laws, which say that energy is not lost; it can only be converted into other forms of energy or matter.

Empedocles had the reasonably correct idea that everything could be derived from four elements, for we should "hear first the four roots of all things: bright Zeus, life-bringing Hera, Aidoneus, and Nestis, who waters with her tears the mortal fountains." More directly, he says:

Come and I will tell you . . .
from which all the things we now see come to be:
earth and the billowy sea and the damp air
and the Titan ether, binding everything in a circle.

If the elementary substances of the universe are only earth, air, fire, and water, then how is it that other substances appear to be different from those four? Once again, we are told not to trust appearances. Should we trust our senses or rely on our ability to reason? The problem of divisibility is central to the problem of trusting the senses. Heraclitus, nicknamed "The Riddler," felt that everything is subject to change and was the first philosopher to profess a distinction between mind and sense.

"It's one thing for the eyes and ears to witness sound and sight," he would say, "but what good are they, if the mind cannot interpret what they hear and see?"

Do we obtain knowledge of nature through reason alone, or do we acquire it through sense alone?

Parmenides felt that we only perceive change through reason. For him, one is persuaded by the virtues of experience, intuition, and compelling forces suggesting that things could not be otherwise. He was referring to this kind of persuasion in his poem, *The Way of Truth*.

The only ways of enquiry that can be thought of:
the one way, that it *is* and cannot *not-be*,
is the path of Persuasion, for it attends upon Truth.

For him, knowledge of nature was based exclusively on reason, which in his time was a newly defined activity, and not, as the Pythagorean experimentalists had supposed, based on observation. "Engrained habit and experience may tempt the use of the blind eye, echoing ear and tongue as instruments of knowledge, but let reason be the test," he would say. "Beware of the senses."

Heraclitus, too, was occupied with the question of which of the two, observation or reason, was the way of truth. For him it was observation, "Because," he would say, "everything changes. So how could reason, which must be fixed, lead to truth about a world where everything changes from one moment to the next?" Not a bad argument, but Parmenides would attack it and ask, "Then how does Earth change to Water or Water to Vapor? Water is less dense than Earth and Vapor less dense than Water. To change from one to the other empty space must be introduced. But empty space is nothingness, which does not exist. Hence there is no such thing as change. The world is one spherically solid motionless universe, incapable of change by the argument that nothingness cannot be something."

Reason had become a new game, complete with that wonderful new logical principle, *contradiction*—after all, a thing cannot be and not be at the same time, just as nothingness cannot be the thing that makes vapor from water. It was a game that would spur intellectual thought over hundreds and now thousands of years to the heights of scientific knowledge.

ZENO ARGUED THAT movement is impossible because in order for a body to move any distance it must first get to half the distance, then half the remaining distance, and so on, forever reaching half of some remaining distance—hence, never reaching the full distance. Aristotle wrote that this paradox suggests that movement is "impossible because, however near the mobile is to any given point, it will always have to cover the half, and then the half of that, and so on without limit before it gets there." Zeno wrote all this in a book, which he claimed was stolen, and which is reported to have contained "forty different paradoxes following from the assumption of plurality and motion." How devastating his loss must have been, writing day after day on scrolls of papyrus, planning ahead, and anticipating each new thought before cutting the skin and sewing in new patches.

There are many variations on this argument, and surely Zeno had considered them. It means any task can never be finished, for in order for it to finish, half the task must be done, and when that is accomplished, half the remaining task must be finished, and so on ad infinitum. The task is general: anything from reading this book to winning gold (a hundred amphorae of olive oil) in a Great Panathenaea chariot race. Mathematicians may simply deny the paradox by claiming that the sum $1/2 + 1/4 + 1/8 + \ldots$ is equal to 1, but they cannot answer the question of *how* the task is actually completed in reality. Mathematics tells us that it happens without explaining why.

At some point after reading his treatise in Athens, Zeno left the Ceramicus with Parmenides, Pythodorus, Aristoteles, and Socrates to retreat to the home of Pythodorus. They walked through a courtyard, through stables, up a few steps to a porch, then through the women's quarters and into a long room with cushioned seats around the walls facing a central hearth over a stone floor. It is likely they encountered preparations for a symposium that would happen late in the evening after the Panathenaea festival—oil lamps being filled, as well as large urns for wine and water.

Here Zeno argued that if one shot an arrow at a target, then examined it at any fixed instant of time, the arrow would appear stationary. If it is stationary at any instant, how can it be in motion? How can it ever even leave the bow, let alone move through the air and reach its target?

One may argue that the very notion of fixing a point in time is absurd and that it makes no sense to say "an arrow appears stationary at any point in time." But in mathematics, time is a variable that can be fixed by declaring it to be some number of units of time from some starting time. Mathematical formulas tell us where an arrow is at any time t, so if we let t equal some specific time, say two seconds after leaving the bow, we should know the exact spot where the arrow is when $t = 2$. But is there any such thing as exactly two seconds, or even an exact spot? We know that if we really try to take a picture of the arrow when $t = 2$, we must have the shutter open for an entire interval of time surrounding $t = 2$. The shutter cannot open and close at the same instant.

Mathematical representations of physics are models that are constructed in the mind. The key to understanding Zeno's arguments is to understand the connection between what it means, both mathematically and physically, to let the time variable be equal to a constant. The mathematician is the conjuror here. Stop time to see the arrow stationary? Yes, that would, indeed, seem to disturb movement, but what we see is not the real arrow; it is another arrow moving in the mind.

Continuity suggests an *uninterrupted* path. We move from here to there without passing through gaps in space. To us, motion seems uninterrupted. Yet, we envision objects moving through space on a line or curve made from an aggregate of points representing numbers, perhaps the distance from one end of the curve. For any number on a number line there is no such thing as a *next* number. So, how do we move from one point to the next, if there is no such thing as a next point? This is the salient arrow in Zeno's quiver. If a path is an aggregate of points, then an object's motion cannot generate a path.

Tobias Dantzig, the twentieth-century author of several popular books on mathematics, put it beautifully: "When we see a ball in flight we perceive the motion as a whole and not as a succession of infinitesimal jumps. But neither is a mathematical line the true, or even the fair, representation of a wire. Man has for so long been trained in using these fictions that he has come to prefer the substitute to the genuine article."

And that's just it. We have been trained in using fictions. We see a ball in flight and presume that what we see is what actually happens. But the mind, not the eye, is the seeing organ. Consider the zoetrope, that nineteenth-century parlor-room toy, in which no more than a dozen still images of a man in various anatomical positions give the illusion that the man is running.

The films we watch are more advanced illusions of continuity. A one-hour film is composed of 86,400 individual still images, yet we see the scenes pass by with utter smoothness. The seventy-two still images on film of a ball in flight for three seconds may look just the same as the real ball in flight. Doubling the number of still images and doubling the speed of the film may not give the viewer any more realistic sense of continuity. There is something biologically magical in that threshold number of frames per second (twenty-four) that tricks the mind into thinking that what we are seeing is continuous. But the mind seems to be able to process far more than twenty-four frames per second, integrating information faster than a film can deliver.

Perhaps there is a good motive for Zeno's motion arguments. Perhaps physical motion simply cannot be represented by mathematical space and time under arbitrarily small intervals beyond measurable experience. The great nineteenth-century mathematicians David Hilbert and Paul Bernays put forward a disturbing answer:

> Actually there is also a much more radical solution of the paradox. This consists in the consideration that we are by no means obliged to believe that the mathematical space-time representation of motion is physically significant for arbitrarily small space and time intervals; but rather have every basis to suppose that that mathematical model extrapolates the facts of a certain realm of experience, namely the motions within the orders of magnitude hitherto accessible to our observation. . . .

Zeno was known as "the two-tongued Zeno" because he often argued both sides of his own arguments, which usually involved either the infinite or the infinitesimal. Two of his paradoxes assume that space and time consist of a finite number of points and instants, while two others make the opposite assumption. There are only three ways out of these paradoxes: either we agree that (1) space and time consists of points and instants, and there are an infinite number of points within any interval; (2) that there are no points and instants in space; or (3) we deny the real existence of space and time altogether.

He was asking such questions more than two millennia before any thoughts of quantum mechanics and relativity, already posing questions contrasting our experiences of motion and our sense of continuity with logical explanations of what we assume to be *reality.* We seem to be comfortable with motion at the macroscopic level by intuiting what we expect to happen through experience, but with no sensory experience at the microscopic level we run into trouble and counterintuitive wonders.

Anyone who believes the atomist argument that all matter consists of atoms and that the atom is indivisible and indestructible must also believe that a moving object must pass from one spot to the next as time passes from one instant to the next. Of course, Zeno was assuming that time moves from past to future through a sequence of successive instants. He was also assuming something far more acceptable: If the object is always moving forward, it cannot be in the same place at two distinct instants of time. We know that Zeno's followers were confused by the meaning of his paradox, but more than twenty-four centuries have passed for intelligent people to have made some sense of it. Even Aristotle seemed to have been confused when he mentioned it in his *Physics.* We now have a clearer understanding of what Zeno could have meant.

Consider three adjacent points labeled *A, B,* and *C.* By this I mean that *B* is immediately to the right of *A* and that *C* is immediately to the right of *B.* In one indivisible instant, an object cannot travel from point *A* to point *C.* If it could, there would be no instant when it could be at point *B.* Of course, this is absurd, because that would mean all motion must take place at the same speed. The only way out of this is to reject the thought that points or instants are consecutive, i.e., arranged in a hierarchy from left to right and vice versa. This leads to equally puzzling thoughts about how a moving body gets from one point to another. If an object moved from A to C, there must have been a moment when it was at a point B between A and C. And there must have been a moment when it was at a point between A and B. This can go on indefinitely.

The stadium paradox asks us to imagine three lines, each either above or below another. Mark the points. The top line has points labeled A_1, A_2, A_3, etc.; the middle line has points labeled B_1, B_2, B_3, etc.; and the bottom line has points labeled C_1, C_2, C_3, etc. The letter indicates the position of the line and the number indicates the position of the point on the line. Now imagine that the lines line up so that the numbers are each above or below each other.

$$A_1 \quad A_2 \quad A_3$$
$$B_1 \quad B_2 \quad B_3$$
$$C_1 \quad C_2 \quad C_3$$

Next, imagine that the top line is stationary, the middle line is moving to the left at a constant speed s, and the lower line is moving to the right at the same speed s.

$$A_1 \quad A_2 \quad A_3$$
$$B_1 \quad B_2 \quad B_3$$
$$\qquad C_1 \quad C_2 \quad C_3$$

Suppose that the line is made up of discrete points. You may have noticed that before these lines moved, A_2, B_2, and C_2 lined up as a column of points, but on the very first instant of movement, the points B_3 and C_1 line up under A_2. It seems that B_3 skipped over C_2 to line up with C_1. In other words, there was never an instant when A_2, B_3 and C_2 lined up as a column. What happened? The answer strikes at Zeno's point. We made one fallacious assumption: that the line is made up of discrete points. We could view Zeno's stadium argument as an indirect proof that the line is not made of discrete points.

Though nature is fantasized as continuous—both by our brains, such as when we are watching a film, and by reason, as argued in Zeno's stadium paradox—she does make jumps. The piece of wood that is divided often enough seems to stay wood for many divisions, but at some point, there will be a specific division when the wood dust suddenly becomes something other than wood. This is the first of several jumps as we continue to split our pieces of matter down to the atom. Eventually, we are left with splitting operations that can take place only in the mind.

· 3 ·

The World Through Aristotle's Eyes

In 343 BCE, Aristotle would take long walks from the palace at Pella to a little gate by the Axius River in Macedonia. He was born in Stagira, a large town near the three fingers of Macedonia jutting into the Aegean, where wild fig trees struggled to grow in rocky soil. Those trees rarely bore fruit, though occasionally someone could find and pluck a lonely fig hidden in their foliage. Aristotle loved to walk, and would often stroll the dusty sandstone road alongside the city wall from the palace to the gate. He was nicknamed "The Peripatetic." Though he was wrong about many details, his gift to the world of knowledge—a contribution that guided the West for more than a thousand years—was an explanation of almost everything.

By the end of his life he had written 337 books on topics ranging from love to medicine. Yet his attire revealed a man calling attention to himself; his clothes were conspicuously fanciful, as were his carefully trimmed hair and the rings on his

fingers. His face was clean-shaven; his body garlanded with or-
naments and jewels. He tutored Alexander the Great in botany,
zoology, and physics. Alexander was only thirteen, and not yet
emperor of Macedonia, Greece, North Africa, Persia, and the
Punjab of India.

Aristotle had a broad concept of *nature*, one that was very
different from the concept we have today. For him, the study of
nature was the study of "all things that move or change, or that
come and go either in some sense of passing from 'here' to
'there,' or in the more extended sense of passing from 'this' to
'that,' which latter phrase is equivalent to 'becoming something
that it was not'—a solid becoming a liquid or a hot thing becoming cold."

The field of change is broad enough to include things that
fall, rise, sink, or expand, and even souls that might transmi-
grate. A stone rolls down a hillside, cold becomes hot, a bubble
is born in boiling water, a block of Pentelic marble becomes the
bust of Hermes, a mind is persuaded by a convincing argument,
or—to paraphrase Aristotle—an uncultivated man becomes
cultivated. These all involve motion in its broadest sense.

Aristotle's theses imply that the cultivation of intelligence
leads to the joys of life. He believed that daily experience and
sensations demand the development of an understanding of
material nature—hopeful and inspiring stuff, after the bleak
Platonic opinion that all knowledge falls short of unattainable
ideals. It may be difficult to imagine a time when all science was
simply thinking about nature, without tests or experiments, a
time when the man on the street could hypothesize about the
universe by feeling that something is true and making a good
argument for its case, a time when there were no laboratories or

statistical samplings to measure probabilities. In the fourth century BCE, reasoning was all that was needed to make a scientific case. Aristotle built his cases from first principles—that is, from indisputable statements—claiming that reasoning is not possible without first principles, definitions, and hypotheses. If he should want to talk of change, he would start by hypothesizing that "wherever anything changes, it always changes either from one thing to another, or from one magnitude to another, or from one quality to another, or from one place to another; but there is nothing that embraces all these kinds of change in common, and is itself neither substantive nor quantitive nor qualitive nor pertaining to any of the other categories. . . ."

Motion for him meant more than just *locomotion*—the movement of an object from one place to another. It meant movement in quality (black to white), or in form (the ripening pear), or quantity (growth in size), or displacement (locomotion). Nature to him was the cause of all things that move, change, or pass from *this* to *that*. "Nature is the principle of movement and change," he wrote. "And since we are interested in Nature, we must understand what 'movement' is. First, we should understand that movement is 'continuous' and that continuity implies the concept of the 'illimitable.'" It was an amazing revelation.

IN HIS BOOK *On Movement,* Aristotle claimed that in order to have movement at all, we must first have continuity, and in order to have continuity we must have division without limit. He was not thinking of division of physical objects such as a stick, which can be divided only up to the point of its atomic indivis-ibles, but of the space and time in which the stick sits.

Anyone reading *On Movement* might ask why something that moves must move through divisible time and space, and the answer is reminiscent of Zeno's: Anything that changes must change in time and space, and hence time must be divisi-ble, for nothing that cannot be divided in time can be made to move in space. Aristotle argued a thing that is undergoing change cannot change from *here* to *there* or from *this* to *that* all at once, for if it did there would have to be an instant when the whole thing became *this* from *that*. He was trying to connect *time* to *change* by making the argument that time is continuous and, therefore, change must be, too.

Aristotle argued for the connection between mathematical continuity and real-world continuity by observing that a travel-ing object cannot skip positions—it must move from one posi-tion to the next. But he was not an atomist. For him, the continuity of space did not imply the infinite division of the ob-ject traveling through space. This seems contradictory, and is reminiscent of Zeno's arguments. How can an object move from one position to the next without space coming in discrete units?

Aristotle wrote, "Movement cannot occur except in relation to place, void and time." He also wrote, "These four things—place, void, movement and time—are universal conditions common to all natural phenomena."

Movement can only happen by direct touch between a mov-ing agent and the moving thing—the stone carver's chisel whit-tles the stone, the potter's hands shape the clay, and the weaver rapidly pushes the weft and shuttle back and forth across a warp through a perfectly synchronized opening and closing heald. For the case of the moving stick, the front pulls the rear or the rear pushes the front.

Aristotle said, "Taking the initiator of movement to mean not that for the sake of which the movement takes place but that which sets it going, we may say that the initiator must be in direct touch with the thing it immediately moves; and by this I mean that there can be nothing between them. This is true of every mover and the moved it directly acts upon."

Hearing involves air particles hitting the eardrum. Seeing involves light waves stimulating the retina. Aristotle could not have known about rods and cones on the retina, and yet, they are in accord with his concept of nature. What about emotions— fear, anger, love? Aristotle attributed those to blood flow. He claimed anger to be "the seething of the blood, or heat in the region of the heart." For him, mind was in the heart, and the eyes were windows to the soul. And all things could be explained by one thing touching and moving another.

Direct contact between the mover and the moved applies to all kinds of motion—locomotion from one place to another, whether the moved is being moved by itself or not; qualitative motion, as in a ripening pear; or quantitative motion, as in the growth or shrinkage of a herd of goats. But anything that moves must move from somewhere to someplace else, or from one state of being to another in some span of time.

But just as motion needs time, time needs motion. In his *Physics*, Aristotle wrote, "So, just as there would be no time if there were no distinction between this 'now' and that 'now,' but it was always the same 'now'; in the same way there appears to be no time between two 'nows' when we fail to distinguish between them." Time and motion are therefore different but inseparable. He asks us to try to imagine time without movement or movement without time. It's impossible. "Even if it were dark and we were conscious of no bodily sensations, but something were 'going on' in our minds, we should, from that very experience, recognize the passage of time." For Aristotle, motion is a gateway into understanding the very fabric of the universe.

TIME IS THE measure of motion—and vice versa. Today we measure time in terms of physical locomotion. Time is simply a recording that separates physical "befores" and "afters." Every moderately precise clock—from Galileo's swinging pendulum to our modern atomic clocks (which oscillate at billions of cycles per second)—measures time by some form of stop-and-go mechanism.

Aristotle presents us with a brainteaser. If all motion were to cease in the universe for an interval of time, what could we possibly mean by that interval? If motion is not taking place, then the time span of the interval is not either; the interval collapses as though there never was one. In other words, every time interval must represent the motion of something in the universe.

There is also a hint of relativity in Aristotle's conception of time. We may ask, What would happen if only one thing in the universe were in motion? We would have to answer that the interval would exist and have some particular measurement, based on the motion of the single moving object. But what would happen to the measure of time when a second object begins to move? Aristotle's answer is that if one object covers less distance in the same time interval than another, then it must be moving "slower" and that time is still the conceptual measure; that is, "we do not speak of time itself as 'swift or slow,' but as consisting of 'many or few' of the units in which it is counted, or as 'long and short' when we regard the continuum . . . for abstract numbers are in no case swift or slow, though the counting of them may be." In effect, he is measuring speeds qualitatively and following a Greek tradition of explaining phenomena through the use of proportions and analogies. Yet we do speak of "swift or slow" as relative terms when we consider distance covered as "great or small" in the time interval considered.

Aristotle believed that if time is continuous, then so is space. Yet time is divided by this curious thing we know as "now"; and, by the same reasoning, so is space. The position of any object in motion is marked and divided by its "now" place in space. But that does not exclude the concept of a smallest unit of time or space. Aristotle surely understood that an interval could be infinitely divided, but his conception of infinity grants that we can always imagine a "beyond"—a potential for continuing indefinitely; that our minds have the power to continue to divide a line or an interval of time as often as we like. But those divisions refer only to rational numbers, the only measurements Aristotle would have known about.

Aristotle uses this potential infinity to argue that Zeno's dichotomy paradox—the argument that a moving object must repeatedly pass a succession of halfway points before getting to its end position—is based on the false belief that it is impossible for a thing to take up an infinite number of positions in a finite amount of time. In effect, a moving object would have to "count" infinitely many numbers before the end of its journey.

Modern mathematics has models that make it possible to perform an infinite number of tasks in a finite amount of time by playing the dichotomy paradox in reverse. David Hilbert's famous infinite hotel trick is a good example: Somewhere in math wonderland there is a hotel with an infinity of rooms numbered 1, 2, 3, and so on. The hotel is always full, but there is always room for one more guest. The manager moves the occupants of room 1 to room 2, the occupants of room 2 to room 3, and so forth. This frees up room 1 for the new arrival. This may seem impossible to accomplish in a finite amount of time, given that the occupants must move in real space and real time. But if the occupant in the first room takes 1/2 hour to move, the occupant in room 2 takes 1/4 hour, and the occupant in the n-th room takes $1/2^n$ hour, then the infinity of moves will be finished in just one hour.

However, Aristotle claims that Zeno had made false assumptions in asserting that it is impossible for a thing to take up an infinite number of positions in a finite amount of time. He points out that time and space are *equally* divisible without limit and therefore there should be not be any surprise that a person can pass through an infinite number of positions in an infinite collection of instants. But there is more to his refutation of Zeno's dichotomy paradox. He claims that when the path of motion is bisected, the motion is interrupted; the bisected point is considered twice—once at the end of the first segment and again at the beginning of next segment.

Modern topology—the branch of mathematics concerned

with special properties that are independent of distance measurement—would be disturbed about this, for it would assume that the point of division lies in one segment or the other, but not in both. So here is Aristotle's argument. If time is continuous and the points of time are represented as points of space, then the point's position must be represented by both the past and future. He argues that Zeno is presuming that if a white object were changing to not-white in a period of time divided into two intervals—A, during which it is white, and B, during which it is non-white—then there must be some instant C when it is both white and non-white; in other words, we are left with the devilishly perplexing contradiction that C belongs to both A and B.

Aristotle argues that the contradiction is based on something he doesn't believe is true: the Pythagorean notion that time is a string of atomic moments, one following directly from its neighbor with nothing in between. This awareness of the nature of number density is significant—it was not fully appreciated by mathematicians before the seventeenth century and the invention of calculus, which depends on the density of irrational numbers in the set of real numbers.

Aristotle argues that if something is moving at one instant it must have already been moving, though perhaps slower or faster. If space and time are both continuous, without "next" points or atomic moments, and if time is merely an intangible numerical scale in our consciousness representing motion, then time is a continuous measure of change in position. It follows that there is no change in position in any instant of time, but it does not follow that no change is taking place.

His definition of "being at rest" means that from one instant to another entirely different instant, the body in question and all its parts occupy the same place. Moreover, he asserts that time is indefinitely divisible. Therefore, when Zeno claims that his flying arrow "does not move" at an indivisible instant, Aristotle agrees that it and all its parts occupy the same place at that instant, but that does not mean it is at rest, for, in order to be at rest, it and all its parts must occupy the same place for a period of time. In other words, whatever is in motion changes position as time continuously moves on; it does not matter what is happening in a single instant.

However, Zeno anticipated his refuters and cleverly designed his four paradoxes to trap them between assumptions of divisibility and indivisibility of time and space. The first two (the dichotomy and Achilles) assume that space and time are infinitely divisible while the second two (the arrow and the stadium) make the opposite assumption.

To refute the Achilles paradox, Aristotle reduces it to the dichotomy by correctly noting that it too is a kind of division of space, not by halves (as the dichotomy supposes), but by a ratio of the speeds of the racers. He also correctly notes that Zeno dupes us into focusing on the moments before Achilles overtakes the tortoise by designing the argument as a catching-up question. Yes, Achilles does not overtake the tortoise while the tortoise is ahead, but we tend to forget that the race continues to the finish line, which may or may not be beyond the point where Achilles overtakes the tortoise.

The fourth paradox seems to be the real trap. In effect, a corollary is that all speeds are equal, for if time and space are made from indivisible atomic instants and points respectively, then a body is forced to pass one atom of space in one atom of time. If that were not the case, then the body would have to pass one atom of space in more (or less) than one atom of time, which would make the atom of time divisible. But Aristotle seems to have misunderstood the point. His brief criticism simply attacks the hypothesis when he says, "The fallacy lies in his assuming that a moving object takes an equal time in passing another object equal in dimensions to itself, whether that other object is stationary or in motion; which assumption is false."

All these arguments seemed to center on the possibility of motion and whether or not time and space were continuous. Cause was a different question. And Aristotle argued that all motion is caused by an external agent, but avoids the question of how that agent continues to do its thing when not in contact with the thing being moved. "If a thing is in motion it is, of necessity, being kept in motion by something." What is that something? His answer is that it is either something within the moving object that keeps it moving, or some other moving agent in contact with it. In his view, motion must be started by something that is already moving and that motion continues only by contact with something that continues to push or pull. The image here is an infinite succession of agents each being pushed or pulled by its neighbor. The idea that a body in motion will continue in motion unless acted upon would have inverted his understanding of cause. He had no concept of inertia the way we do to explain why a stone continues to travel after it leaves the hand that throws it. That concept was still a millennium away.

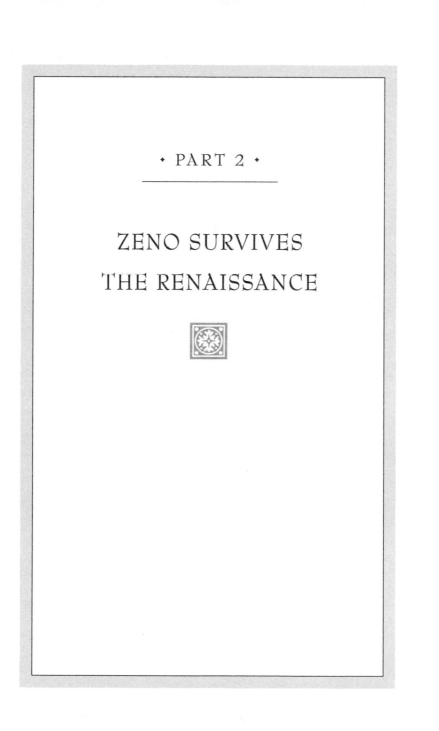

· PART 2 ·

ZENO SURVIVES
THE RENAISSANCE

· 4 ·

Speed Becomes a Quantity

If in 1265 we turned to Thomas Aquinas, the Italian Dominican friar, theologian, philosopher, and most influential sage of medieval times, and asked "What causes movement?" his answer would have been a simple one, perhaps as curt as, "God." We would find Aquinas somewhere in the Kingdom of the Two Sicilies (present day Naples), perhaps sitting on gently sloping hills above ripening grapes in vineyards extending to the limits of vision, contemplating heaven.

"Reason and faith," he might say, "are not contradictory. They are both gifts from God and may be reconciled to discover and prove His existence."

A dozen years later, Aquinas's answer to our hypothetical question was actually made official by a papal decree designed to suppress all contradictions to church teaching. It announced that Aristotle and the Arabs were infidels and declared God as the maker of motion. But by then the church was too late. The Crusades were ending, bringing back to Europe intellectual treasures from highly developed civilizations from Persia to Libya.

For a thousand years, from about when the Visigoths ripped through Constantinople, overran Greece and sacked Rome, Europe's intellectual growth had been anesthetized by Christian faith and Church dogma. In 392, a year after the Emperor Theodosius I issued an edict declaring paganism "a crime of high treason against the state, which can be expiated only by the death of the guilty," Christian marauders torched the library of the pagan Temple of Serapis (the Serapeum), which contained more than 300,000 scrolls, and murdered several of the Museum's scholars on the streets of Alexandria, including Hypatia, a woman mathematician. By the seventh century, hundreds of monasteries and hostels lined the roads and ports from Canterbury to Jerusalem, providing a highway of taverns and lodgings relaying information and tourist guidance. Tourism swelled as hostel owners profited. "Then people long to go on pilgrimages / and palmers long to seek the stranger strands / of far-off saints, hallowed in sundry lands / and especially, from every shire's end. / In England, down to Canterbury they end / to seek the holy blissful martyr, quick / to give his help to them when they were sick." Some pilgrims left their homes to permanently wander from one holy place to another, but, distinguished by their wide, flat-crowned hats, they soon fell prey to bandits and thugs.

As the unified world under the Roman Empire broke apart, the Islamic world rose. Muslims conquered the south of the Mediterranean from Syria and Mesopotamia to Spain, expanding well beyond the limits of Roman civilization, spreading into Asia and Africa. Arabs brought inventions back from China and India, advanced astronomy, introduced the Hindu notion of zero, invented algebra, developed the chemistry of metallurgy, and invented the mizzenmast to speed their ships. Jerusalem became a prized conquest. In 632 Muhammad died and, as it is said, ascended to heaven from the rock at the Temple Mount (also the site of King Solomon's Temple).

On November 27, 1095, Pope Urban II addressed a large crowd in a wheat field in Clermont, France. "Jerusalem is the navel of the world," he called out. "A land which is more fruitful than any other, a land which is like another paradise of delights. This is the land which the Redeemer of mankind illuminated by his coming, adorned by his life, consecrated by his passion, redeemed by his death and sealed by his burial." And with a passionate plea, he incited the crowd to take up arms against all heathens. "This royal city, situated in the middle of the world," he continued, "is now held captive by his enemies and is made a servant, by those who know not God, for the ceremonies of the heathen. It looks and hopes for freedom; it begs unceasingly that you will come to its aid. It looks for help from you, especially, because God has bestowed glory in arms upon you more than on any other nation. Undertake this journey, therefore, for the remission of your sins, with the assurance of 'glory which cannot fade' in the kingdom of heaven." When the Crusades were over, spoils were brought back to churches and monasteries all over Europe. Among the treasures were silks, perfumes, spices, and books. The books were written in Arabic— translations and transcriptions of Greek and Egyptian scrolls stolen from Arabian libraries.

The teachings of Aristotle were no longer confined to Greece, Egypt, and the Mediterranean. Aristotle's eight books on physics came from lecture notes compiled over many years of meditating and talking about motion and change. He was known for his work in logic, but his works on physics and motion were emerging just when universities began to open all over Europe.

Aristotle's works were banned at the University of Paris. Only theologians were permitted access, though many others could read Aristotle in private. Heresy was a serious crime, and any unauthorized person found reading Aristotle would be considered a heretic and imprisoned for life. A copy of a proclamation written by the provincial synod of Sens and signed by the Bishop of Paris in 1210 still exists:

> Let the body of master Amaury be removed from the cemetery and cast into unconsecrated ground, and the same be excommunicated by all the churches of the entire privince. Bernard, William of Arria the goldsmith, Stephen priest of Old Corbiel, Stephen priest of Cella, John priest of Occines, master William of Poitiers, Dudo the priest, Dominicus de Triangulo, Odo and Elinans clerks of St. Cloud—these are to be degraded and left to the secular arm. Urricus priest of Lauriac and Peter of St. Cloud, now a monk of St. Denis, Duarinus priest of Corbiel, and Stephen the clerk are to be degraded and imprisoned for life.
>
> Neither the books of Aristotle on natural philosophy nor their commentaries are to be read at Paris in public or secret, and this we forbid under penalty of excommunication.

When the University of Paris was shut down in 1229 because of a dispute between the university and the local authorities, the newly established University of Toulouse found its chance to lure Paris students and masters. Its representatives distributed fliers that read: "Those who wish to scrutinize the bosom of nature to the inmost can hear the books of Aristotle which were forbidden at Paris." Papal decree suppressing all contradictions to church teaching was too late.

Thomas Aquinas was in residence at the papal court in Orvieto near Rome in 1262. Pope Urban IV was greatly interested in philosophy and surrounded himself with a number of talented scholars and philosophers. There, Aquinas met William of Moerbeke, who had translated several works of Aristotle from Greek to Latin along with his own commentaries. This gathering of great intellectual talent inspired works to make Aristotle more available to European scholarship. This was when Aquinas set to work writing his commentaries on Aristotle. Though Aristotle was a pagan, Aquinas's commentaries turned an otherwise obscure *Physics* into a clear and brilliant explanation of what Aristotle had in mind. Later, in the twentieth century, the Aquinas scholar Vernon Bourke would go so far as to say, "It is a clear presentation of the sort of cosmology from which men like Copernicus, Galileo, Kepler, and even Newton took their start in founding modern astronomy and physics."

BY THE TIME the fifteen-year-old king, Edward III, took the throne of England in 1328, universities had been chartered and well established in Oxford, Cambridge, Paris, Toulouse, Padua, and Naples. All degrees and teachers had to be approved by one of the popes, a control inherited from the time when the university was a guild of teachers and students of the cathedral schools.

By then Aristotle's works were permitted and fashionable. Aquinas's commentaries and interpretations had made them acceptable to the church. The *Physics* was the wisest thing available and—though it had no references or glories to a Christian God—it did not seem to interfere with church teachings. So, fourteenth-century physics was mostly Aristotelian, describing motion as conditioned by time, comparing "velocities" as one being quicker than another. But the measure of velocity—as a quantity of something involving space and time—did not come about until the thirteenth century.

The third-century BCE mathematician Autolycus tried to define uniform velocity of an object by casting off all its unessential matter and considering the object as a point moving equal distances in equal times. This is a purely geometric definition, idealized by points and lines.

"The velocity of a point," he said, "is uniform when that point traverses equal linear distances in equal periods of time." This means that for uniform velocity the ratios of the distances traveled by the moving object equals the ratio of the times it takes to travel those distances. We, in the twenty-first century, would think that speed is determined by the ratio of distance to time. But this would have been a problem for a medieval physicist who took too much direction from Greek authors such as Autolycus, who thought ratios had to be between like units, distance-to-distance and time-to-time.

Gerard of Brussels came a step closer to defining velocity as a ratio of two unlike quantities such as distance and time. We know next to nothing about Gerard, except that he wrote the first Latin treatise on kinematics, that branch of the study of the dynamics of motion that deals with aspects of position, velocity, and acceleration without regard to mass or force. We know that he was instrumental in reviving the mathematical works of Euclid and Archimedes. We have a fragment of his book *Liber de motu* (*Book on Motion*), but we don't even know in which century the book was written. The best guess is that it was written between 1187 and 1260.

Gerard said, "The proportion of the movements (i.e., speeds) of point is that of the lines described in the same time." This short sentence is responsible for impressive breakthroughs in kinematics that would occur a century later. Until Gerard made this statement, everyone assumed that uniform speeds were proportional relationships between spaces and times. In other words, motion was talked about as proportions relating distances to each other or times to each other, but never as a comparison between space and time. Gerard could compare velocities by comparing distances traversed in equal times. This may seem terribly odd to us, who know velocity only as a ratio of space to time. Here, for the first time, someone is treating velocities as magnitudes, sparking a shift toward the modern view of instantaneous velocity, a raw ingredient that would incubate and germinate for another 400 years in wait for calculus—a subject that thoroughly altered the study of motion by using infinity to model how things change with time. Gerard's insightful idea was a start, but another step had to be taken.

IN THE BEGINNING of the fourteenth century, the causes of motion were still not well understood. *Omne quod movetur ab alio movetur* (Whatever is moved is moved by another) was still an acceptable aphorism.

Sometime between 1328 and 1350, at the newly established Merton College, Oxford, a new idea was emerging. Like most medieval cities, Oxford was a walled town in 1328. If you walked southeast on the cobblestone street just inside the wall toward Saint John's Lane you would come to the Church of Saint John the Baptist, a few stone manor houses, and a three-story building built from Cotswold stone, a yellow limestone that was beginning to turn the color of honey. Its steep-pitched, dormerless stone roof provided the building with a third floor of dormitory space and a library at the east end where another building adjoined it at right angles. The first floor had one large room with windows facing south onto a small lawn without plants or trees. This lawn would later become a quadrangle in the 1370s after two more buildings were built, a model not only for other Oxford colleges but also for colleges and universities throughout the Western world.

In these buildings something unusual happened sometime between the years 1328 and 1350. Four mathematicians from Merton worked together to bring forth the first breakthrough on measuring acceleration.

Thomas Bradwardine, William Heytesbury, Richard Swineshead, and John Dumbleton worked on an idea that changed the world. Bradwardine, known as "doctor profundus," was clearly the senior of the group. He had just completed his *Tractatus de Proportionibus Velocitatum*, a book about kinematic problems that was a strong but unanticipated influence on what was about to happen at Merton. He lectured on the causes of motion, though he didn't exactly know those causes before leaving Merton for the royal court at Flanders, after which he became chancellor of St. Paul's Cathedral and later the Archbishop of Canterbury. He held that last position for just one year before he died of the plague.

With gunpowder and firearms appearing in Europe, it wasn't too long before the first cannon was cast to threaten the era of armored knights in fortified castles. The first cannon was probably fired at just about the time that the four Merton College mathematicians were sharing their ideas on the mechanics of motion. They did not fire cannonballs, but rather arrows at the ends of bolts.

For the first time in history, the causes and effects of motion were beginning to be distinguished and understood. This was the moment when the ideas of instantaneous velocity and uniformly accelerated motion were emerging to set the stage for what was to become (300 years later) one of the motivating applications of calculus. It was also the moment of a serendipitous discovery linking acceleration to distance for a freely falling object.

Gerard of Brussels gave the jumping-off point for the Merton treatises. Now Heytesbury was delivering one of his lectures on motion. He talked about what has become known as the *acceleration theorem*, a theorem that applies to freely falling objects, which are assumed to accelerate uniformly. Under that assumption, in each and every increment of time, the object acquires an equal increment of velocity. In other words, the object moves twice as fast at the end of the second second than in the first, three times as fast after the third second than in the first second, and so on.

Uniform acceleration means that speed is increasing at a constant rate; so Heytesbury argued that if the object starts falling from rest, at the end of any time interval the distance traveled must be the product of the average speed and elapsed time. If, at the end of 2 seconds, the object's speed is, say, 64 feet per second, then it would have traveled (32 feet per second) × (2 seconds) = 64 feet. At the end of the first second, it would have traveled (16 feet per second) × (1 second) = 16 feet. He noted that in this one example, the object falls three times as far in the second second than in the first. Extending this example, he found that at the end of 4 seconds, the object's speed is 128 feet per second, and in that time, it would have traveled 64 × 4 = 256 feet.

Heytesbury noticed that at the end of two seconds, the object would travel four times the distance that it would in one second. In four seconds, the object would travel four times the distance that it would in two. He thought that it could be a rule; perhaps the object always travels four times as far when the time interval is doubled.

If his hypothesis was correct, then he had hit on something truly magnificent, for the distance traveled by the end of 2t seconds may be computed by realizing that the object must be moving at twice the average speed for twice the time. And therefore, the object always falls four times as far when the time interval doubles. The only way this could happen is if the distance traveled is proportional to the square of the time of travel. The respective distances traveled in 1, 2, 3, 4, . . . seconds are the squares 1, 4, 9, 16, ... In each second, the distances increase as the series of odd numbers 1, 3, 5, 7, ...

This says that the final velocity of the object is twice the ratio of distance to time of fall. And what if the object does not start from rest, but has an initial velocity v_0 downward? Then the distance traveled is the average of its initial and final velocity multiplied by the time of travel. This formula is remarkable for three reasons: (1) For the first time in history, an actual number for velocity can be determined from knowing the distance and time, (2) the ratio of distance to time means a ratio of different units, and (3) the formula is accurate, even from today's viewpoint. We know that $v - v_0 = gt$, where g is the acceleration due to gravity, v_0 is the initial velocity, and v is the final velocity. So, when we substitute gt for $v - v_0$ in the formula, we get the same formula that every calculus student of today knows: $s = v_0 + 1/2gt^2$.

This algebraic model of motion offered a superficial escape from the difficulties of Zeno's paradoxes. In general, if Achilles's speed is A miles per hour, the tortoise's speed is B miles per hour and the tortoise is given a head start of H miles, then in t hours Achilles will cover a distance of At miles and the tortoise will cover $Bt + H$ miles. To find the time it will take Achilles to catch up with the tortoise, one had only to solve the equation $At = Bt + H$ to get $t = H/(A - B)$. Note how much can be read from this formula: (1) A must be larger than B, for otherwise time would be negative. (2) If $A = B$, the denominator is zero and the model is invalid. This little model presupposed that Achilles would eventually catch up with the tortoise, permitting the algebraist to equate the distance traveled by Achilles with that of the tortoise. (3) The model assumes there is some way of determining the speed of each racer. And (4), it assumes (just as Gerard did) that speed equals the ratio of distance over time (so that the distance covered by a known speed and known time could be determined).

However, the escape of difficulties is superficial because the algebra avoids any mention of a leading phenomenological concern; that is, *how* Achilles overtakes the tortoise.

Galileo Galilei, the Father of Modern Science

On a Sunday morning in 1583, young Galileo Galilei would have walked along the cobblestone quay of Santa Maria della Spina to attend mass at Pisa Cathedral. Other parishioners would have filled the streets as bells from small churches accompanied the loud gongs coming from the leaning tower beside the cathedral. Walking along the banks of the Arno, crossing Brunelleschi's magnificent Ponte a Mare, Galileo would have seen the masts of fishing boats slowly swaying against a backdrop of stone houses, like metronomes to the regular beat of the rippling river and ringing bells.

He was living in the house of a relative near the Porta Fiorentina. It was a short walk from there to the cathedral, but long enough for the talented student to spot synchronization of sound with scene. Flags atop masts waving and flapping to the strict rhythm of rocking boat decks. He might have noticed that the time it took for a mast to complete its swing did not change, an observation that could have incubated in the subconscious, waiting for important connections before hatching a scientific discovery.

The cathedral was modest—not very tall and architecturally plain, with a facade mimicking that of its leaning bell tower. Inside the nave was a large, ornate, bronze chandelier with thirty candles in three tiers balanced around its periphery. A drip tray hung by three short chains under each candle. To light the candles the giant fixture would be lowered and raised by a chain through its center. If this were done just before mass—as it likely was—the entire fixture would gently oscillate until it calmed down to faint undulations caused by vibrations of the church organ. Most worshippers would hardly notice.

But a genius like Galileo, sitting through a dull sermon, would have wandering thoughts. If a chandelier were swaying, he would watch it and marvel at its motion. One imagines him timing the oscillations against the timing of his own pulse and having revelations about how to measure time. History claims that Galileo timed the chandelier while sitting through a humdrum mass in Pisa Cathedral. If the story were true, he would have noticed a magnificent phenomenon: Though the chandelier would slow down, the duration of each swing would not. Shorter swings would simply be slower. The time required for any complete to-and-fro swing of the chandelier would depend only on length. Surely this is a fictional account of how Galileo made his first independent discovery about motion, because, by other accounts, the cathedral chandelier was not installed until 1588, five years after Galileo claimed the discovery.

True or not, the story tells of how such a discovery does surface. Galileo could have come to the same revelation observing any swinging object. The mathematician Vincenzo Viviani, a pupil and friend of Galileo, perpetuated the story, saying, "Having observed the unerring regularity of the oscillations of this lamp and of other swinging bodies, the idea occurred to him that an instrument might be constructed on this principle, which should mark with accuracy the rate and variation of the pulse." Some say that it was that incident at the church that influenced Galileo to study mathematics; others say it was a mathematics lecture that he attended by chance.

In late-sixteenth-century Italy, professors were still entrenched in Aristotelian doctrine, in the belief that the Greeks had already expressed all worthwhile knowledge, and in a mistrust of new ideas coming from creative minds. A century of unmatched exploration had just passed, more than doubling the size of the known world. America was discovered. Vasco da Gama had sailed around the Cape of Good Hope to reach India. Magellan had sailed clear around the entire world. The vast Pacific was discovered. Europeans had stood on the continent of Antarctica and had not fallen off; stones still fell to ground even on the other side of the planet; and the world was suddenly believed to be spherical and more than just Europe, Asia, Africa, and the Holy Land.

For a thousand years, thinkers had been sitting in their dimly lit studies, university libraries, and secluded monasteries, rationalizing and arguing about the shape of the planet, the makeup of the celestial sphere, or the laws of nature. But it took the courage of Portuguese, Spanish, and Italian sailors venturing into the dangers of vast unknown seas to determine perceptible truth. Nobody would ever again deny that a great sphere of constellations completely surrounds our small spherical planet. The discovery of the Americas offered a new perception of the world. America was not discussed in the Bible. Ptolemy never mentioned it, nor did Aristotle in his *De Caelo*, nor Pliny in his *Natural History*. So, when Spanish conquistadors returned to Europe laden with myriad wonders that had never been seen or written about, some folks at home began to experience exceptional curiosity, questioning why their handy classics never said a word about the existence of such exotic lands of improbable flora and fauna. The conquistadors found delicious fruits, vegetables, and nuts never seen in the Old World—tomatoes, corn, avocados, pineapples, cranberries, blueberries, sunflower seeds, cashews—and cochineal, giving the most intense, concentrated, brilliant red dye the world had ever seen, a dyestuff so precious that it later set off wars and encouraged piracy in the Atlantic from the hidden coves of the Caribbean to the port of Cádiz. Hesitant to defy their Bible, they challenged the wisdom of established classical intellectual teachings and began to investigate nature by direct observation.

And of course, the moveable type of Johannes Gutenberg's printing press in 1436 with replaceable wooden or metal letters and the invention of plant-fiber paper were responsible for the publication of more books in the sixteenth century than had been produced in the 3,500-year period since the first Babylonian author produced the first cuneiform tablet.

Teaching was dictatorial, and rote memorization of Aristotle's works played a central part in the curriculum. The seven liberal arts—grammar, logic, rhetoric, arithmetic, geometry, music, and astronomy—were required, though how much of each was a matter under local control. This rote learning numbed the intellect so severely that nobody thought to criticize the classic works of science, especially the unshakable doctrines of Aristotle. Moreover, except for rote learning of arithmetic and computation, mathematics was completely neglected. "The names of Euclid and Archimedes were empty sounds to the mass of students who daily thronged the academic halls of Bologna, the ancient and the free, of Pisa, and even the learned Padua."

The Italian humanists, who studied the principal literature of antiquity for literary content—as opposed to theological matter—accepted printing with scorn: "Printed books seemed a cheap substitute for their beloved manuscripts, nor did they wish any enlargement of the reading public to include persons without taste. Taste, style, manner, correctitude, *aplomb* were set above more substantial attainments."

But the works of Archimedes, which had been copied into Greek in the ninth century and translated into Latin in the fifteenth, were now being printed and sold throughout Europe. These works were beginning to inspire a new generation of independent thinkers to rethink old doctrines of motion and mathematics.

YOUNG GALILEO WAS studying the usual courses of philosophy and medicine, but under stiflingly rigid training, rather than through the kind of education he was used to at home with his father, who taught him to weigh, examine, and reason the truth of each assertion before accepting it. He despised university training, which professed truth by authority and regarded any contradiction to Aristotle as blasphemy.

His teachers found him obstinate and uncooperative. He secretly, and without tutors, read the first six books of Euclid before convincing his father that he had rare mathematical ability and that he should study mathematics rather than the more lu-crative field of medicine. How fortunate, because from the be-ginning, he studied mathematics with great passion, thinking of it as the means to understand nature's most hidden secrets, to transform scientific observations into sensible and practical principles.

Several years later, after being given the title of Professor of Mathematics, he began to recognize that the study of motion—the concept of motion itself—was central to the scientific understanding of all natural phenomena. He read a book of speculations on mathematics and physics by Giovanni Battista Benedetti, which described a theory established two centuries earlier by Jean Buridan at the University of Paris.

Appropriately called "Parisian physics," the "new" physics assumed that air was not the cause of motion, as Aristotle had, but rather that the object itself contained the cause—the object had "impetus." Though this new theory begged to answer the question of what impetus is, the idea excited Galileo and gave him courage to abandon Aristotle's ideas on motion. However, Galileo did not abandon Aristotle's empirical methods; he merely combined observational methods with mathematical reasoning to put physics on a stable mathematical footing.

In 1590, he wrote in his treatise *De Motu* (*On Motion*), "The method that we shall follow in this treatise will be always to make what is said depend on what was said before, and, if pos-sible, never to assume as true that which requires proof. My teachers of mathematics taught me this method."

He refuted Aristotle's arguments with a style clearly influ-enced by his reading of Euclid and Archimedes. He did exactly what he said he would do—what he would say in one chapter would depend on what he had said in the one before. The book exploits two brilliant, central clues to understanding motion. The first was to use Archimedes's lever principle to compare speeds of heavy objects to light ones in the same medium. The second was to use hydrostatics to compare movements of moving objects of equal weight in different media. He presented his definition of "heaviness" and "lightness," then alleged that heavy things naturally move slower than light things and that natural motion is caused by heaviness or lightness. Later, he proved that bodies of the same heaviness as the medium neither move upward nor downward, and that bodies that are lighter than water could not be completely submerged. He created an analogy between bodies moving naturally and the weights of a balance to get at the cause of speed and the slowness of natural motion, and found that different bodies moving in the same medium maintain a ratio (of their speeds) different from that attributed to them by Aristotle. Everything he said was considered in physical terms, so bodies moving naturally are reduced to the weights of a balance.

Aristotle claimed that two bodies made from the same material would fall at speeds that are proportional to their sizes, so a large piece of gold would fall faster than a small piece.

"How ridiculous this view is, is clearer than daylight," wrote Galileo. He then gave several salient examples to damage Aristotle's view. But his most striking example is a logical observation. He argued that if two bodies of the same material and weight were let go in a medium, then Aristotle would be forced to say that the two bodies together would descend faster than either one alone.

"What clearer proof do we need of the error of Aristotle's

opinion? And who, I ask, will not recognize the truth at once, if he looks at the matter simply and naturally?" Galileo presented it so simply and naturally that one wonders how it was possible for Aristotle to have missed Galileo's argument. Just take the extreme case where one object is a thousand times heavier than another. Galileo used such extremes to ridicule Aristotle's defenders. He wrote, "Surely, these people must do some toiling and sweating before they can show that the velocity of one is a thousand times that of the other." Galileo was on a path to a point of great consequence; he was about to discover a marvelous property of mathematics together with an ingenious model of physics. He was not only about to discover something astonishing and in full contradiction to Aristotle's belief, but he was to do so in a most inventive way. Aristotle said that speeds of bodies falling in different media are in proportion to the rareness of the media. Galileo wrote, "These are Aristotle's words, but surely they embrace a false viewpoint."

First he had to clarify what he or Aristotle meant by the speed of a freely falling body. Galileo said that a freely falling body accelerates, causing the speed to change at every moment. So what could he have meant by the speed of a falling body? We can only assume that he meant the speed after acceleration has ceased, that is, when the body has come to its maximum speed in the medium.

"And to make this perfectly clear," he continues, "I shall construct the following proof."

His proof may be paraphrased as follows: Suppose that the rareness of water is 4 and that of air is 16. Take a body that does not sink in water, say wood. Suppose its velocity in air is 8. Its velocity in water is 0 because it doesn't sink. Surely there is some medium such that its speed is 1. Call that medium X. Since the body moves faster in air than in water, the rareness of X must be less than 16. Aristotle would say that the ratio of rareness of media must equal the ratio of speeds. But that means that we have the analogy X is to 16 as 1 is to 8, which means that $\frac{X}{16} = \frac{1}{8}$; therefore $X = 2$, and so the rareness of X must be 2. But how can our piece of wood float in water and sink in a medium whose rareness is less than water?

Galileo jibed, "Can anyone fail to see the error in Aristotle's opinion?" and went on to tell us what the true ratio is. "Take an amount of each medium equal to the volume of the body, and subtract from the weights [of the amounts] of each medium the weight of the body. The numbers found as remainders will be to each other as the speeds of the motions." In other words the weights must be taken relative to the medium. So, for example, take aluminum. One cubic centimeter of aluminum weighs 2.6 grams in air, but only 1.6 grams in water. Galileo wanted to use weights relative to the weights of the media in computing the ratios of velocities in two different media.

He destroyed Aristotle's principles of physics, one by one, and argued proofs for new principles. Once again, he jibed, "Aristotle, as in practically everything that he wrote about loco-motion, wrote the opposite of the truth on this question, too. And surely this is not strange. For who can arrive at true conclusions from false assumptions?" Eventually, he came to the question, "By what agency are projectiles moved?"

It is a very challenging question, whose answer did not come until after Isaac Newton announced his law of inertia, which says that an object will not change its state of motion unless it is forced to. The Latin meaning of "inertia" is, in effect, *laziness*. And the word is used to imply that inertia causes an object to lazily continue to do what it is doing. So, in the absence of any forces, a moving object will maintain its direction and speed. This answers Galileo's question; inertia keeps the projectile moving. This may seem like dodging the question, relegating the answer to some phenomenological embodiment that the object absorbs and maintains until some external event takes it away. Those of us who have grown up believing in inertia see no problem with inertia as the answer to Galileo's question. We simply say nature, through its laws, acts on the object.

Aristotle's answer was that a thrown stone sets continuous parts of air in motion, which move other parts in succession. When the stone is released, it moves along by those portions of moving air. Galileo wrote that "Aristotle and his followers, who could not persuade themselves that a body could be moved by a force impressed upon it, or recognize what that force was, tried to take refuge in this view." He then demolished Aristotle's view by giving several compelling examples. How does the arrow, shot from a bow, move so swiftly against a strong wind? Aristotle's followers would be forced to say that the wind blows against itself.

Once again, Galileo remarked, "They are not ashamed to utter such childishness." Or take a ship propelled by oars against the current. How does it move when the oars are taken out of the water? "Who is so blind as not to see that the water actually flows with very great force in the direction opposite to that of the ship?" he asked.

In another argument he asked his followers to consider a perfectly smooth spherical marble that can rotate on an axis through its center. Spin the marble and it will continue to rotate. But the surrounding air is not moving, for there is nothing to move it.

And for what he considered his most beautiful example he asked us to think about what passes from a hammer to the bell of a church tower when the hammer strikes the bell. Both hammer and bell are silent before the strike. But after the strike, the loud sound comes from the bell and continues for quite a while after the hammer is pulled away—and gradually diminishes. "But who of sound mind will say that it is the air that continues to strike the bell?" he asked. "If it is the air that strikes the bell and causes the sound in it, why is the bell silent even if the strongest wind is blowing?" he asked. "Can it be that the strong south wind, which churns up the whole sea and topples towers and walls strikes [the bell] more gently than does the hammer, which hardly moves?" He did not fully answer the question, but came as close as he could have for his time. He simply said that projectiles move by a driving force given by the thrower.

The Aristotelian doctrines of motion began to crumble as more and more scientists and natural philosophers were basing judgments on real-world experiments rather than purely intellectual reasoning. Inconsistencies popped up with each new experiment. Each new inconsistency was met with a tailoring of meaning. "Well, Aristotle meant to say..." his supporters would say, until a blitz of discrepancies forced too many unnatural alterations into a quilt of conflicting patches of truth.

ALTHOUGH GALILEO IS often credited with experimentally debunking Aristotle's claim that heavier objects move faster in free fall, several other stories suggest that others had performed similar experiments before Galileo—as legend has it—dropped two objects of different weights from the Leaning Tower of Pisa. Some claim that others carried out the test as early as 1544. Another claims that Galileo's predecessor at the University of Padua performed the experiment in 1576. However, the more likely fact is that it was Simon Stevin who performed the experiment in 1586, not at the Leaning Tower of Pisa, but from another leaning tower a thousand kilometers from Pisa.

Delft was a small walled town in the southwest Netherlands. Sometime between 1325 and 1350, a clock tower was constructed alongside a small thirteenth-century parish church. The tower was built on fill, and like the tower in Pisa, after construction was completed, began to lean considerably to the northwest. Perhaps the lean intensified after a magnificent nine-ton bell was installed just fifteen years before Stevin climbed the tower to perform his experiment. The bell's sound caused enough acute vibrations to severely damage the tower, so it was rung only on special occasions.

One can only imagine the technical challenge of lifting a nine-ton bronze bell to the top of the tower and securing it to the oak bell cage. Stevin might have been around to witness the lifting and would have thought it a marvel, crediting the basic idea of mechanical advantage to his hero Archimedes. He may have been inspired to work on the science of mechanics and statics, and to make important improvements on Archimedes's work. He spent the year 1586 working on a theorem about the triangle of forces, not knowing that the theorem would revolutionize the way scientists look at forces and reform the science of statics.

His theorem may have come from a popular puzzle of the time. The puzzle came from a thought experiment involving a perpetual-motion machine that was discussed among students at Leiden University when Stevin was a student there. He discussed the puzzle with his good friend the young prince Maurice of Orange, son of William of Orange, the lieutenant governor of the Netherlands who was assassinated while leading a revolt against the Spanish.

Students often met at a popular basement tavern near the university to discuss what they considered intellectual thought experiments. Water, dripping from cracks in its massive stone walls, kept the tavern cool and damp. Candles and torch sconces provided moderate light in the windowless room. An intoxicating smell of fermenting spirits seeped from a whiskey and brandy distillery next door. Beer was cheap. Stevin, Maurice, and other friends would often sit together at a long sticky oak table coated with layers of sugars dried from decades of beer spills to drink and pose challenging riddles. Some were easily solved, but the one they returned to day after day—the "hanging chain mystery"—was truly daunting.

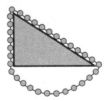

A pearl necklace hangs loosely over a right-triangular wedge of height 3 meters, base 4 meters, and slope 5 meters. The entire system is (absurdly) assumed to be almost frictionless. If the necklace has uniform density of d pounds per meter throughout, then the weight of that part that rests on the slope is $5d$ and the weight of that part that hangs straight down is $3d$. In this situation, the weight of necklace over the sloping section of the wedge is much greater than the weight hanging straight down. Maurice claimed that the necklace should roll down the slope by virtue of the larger weight on the slope. He argued that the loop under the wedge is fully symmetric and therefore could not contribute movement one way or another. However, if the necklace should start to roll down the slope, it should perpetually continue to do so because its movement does not change the condition of which side has more weight.

Stevin's study of this bizarre situation eventually led him to the idea that the surface of the wedge itself affects the size of the gravity force. His answer was that the slope of the wedge diminishes the downward force of gravity, conveniently splitting that force in two—one in the direction of the slope, the other perpendicular to the slope. The second force is totally annihilated by another force perpendicular to the slope, the one exerted by the sloping floor of the wedge. In effect, the only true force that could produce motion would be the force along the slope. That force is only 3/5 of the weight of that portion of the necklace resting on the slope, which is $(3/5)5d$. But $(3/5)5d = 3d$, which is precisely the downward force exerted by the vertical 3 meters of necklace. The forces are equal, the necklace is in static equilibrium, and it will not move, even in the absence of friction.

By decomposing the forces in this interesting system, Stevin introduced the principle of virtual force, the principle that guided the science of mechanics through four centuries to the present day. It was an early application of frictionless Platonic ideal geometric thought to the real world of mechanics, a first step in the journey to a complete understanding of the mechanics of motion.

What Stevin did was to idealize the real world of mechanics. He broke with the traditional existential imagery in discussions of weight, force, and motion to bring in the help of the conceptual world of mathematics with its idealized Platonic interpretations of the essentials through perfect points, lines, space, and the geometrification of forces, which until then were considered quantities. James Newman addressed this geometrification in his commentary on Galileo.

"The ghosts of Plato and Pythagoras," he said, "returned triumphantly to point the way. Modern mechanics describes quite well how real bodies behave in the real world; its principles and laws are derived, however, from a nonexistent conceptual world of pure, clean, empty, boundless Euclidean space, in which perfect geometric bodies execute perfect geometrical figures."

EVERYONE SHOULD READ at least a small part of Galileo's *Dialogues Concerning Two New Sciences* to get a formidable impression of the author's brilliance. According to Galileo himself, they contain the most important results of all his studies. The book is written in play form with three characters, Salviati, Sagredo, and Simplicio, in conversation over a period of four days. Day three is about the motion of falling objects. Through a magnificently organized set of axioms, Galileo—in the voice of Salviati—manages to elegantly prove a remarkable fact. Take two frictionless inclined planes of the same height but different slopes. Roll the same object down each of the planes. When it reaches the bottom, its speed will be the same for each plane. In other words, the speed at the bottom of an inclined plane depends only on the vertical distance of the starting height and is independent of the slope. This is a significant proposition that must have astonished everyone thinking about motion.

Another observation hinted that conservation of energy comes from an experiment anyone can do. Take a string, say six feet long, and tie a small weight—say, one pound—to one end. Tie the other end to a nail in the wall at some height—say, eight feet—from the floor. Pull the weight to one side keeping the string taut so it rises to four feet from the floor. Ignoring friction, the pendulum will swing back and forth with the weight always rising to the same height. Now place another nail at any height between two feet and four feet from the floor and on the vertical line below the first nail. Again, pull the weight to one side keeping the string taught so it rises to a height four feet from the floor. When the weight swings past the lower nail, the string will hit the nail to prevent the full swing of the pendulum, but the weight will still rise to the same height of four feet, as though the nail were not there.

He investigated naturally accelerated motion simply and easily, for, he thought, "no one believes that swimming or flying can be accomplished in a manner simpler or easier than that instinctively employed by fishes and birds." So when a stone falls from rest, it must fall in a manner that is exceedingly simple. "If now we examine the matter carefully," he observed, "we find no addition or increment more simple than that which repeats itself always in the same manner."

What an extraordinary observation. He learned not only the truth about freely falling objects, but also a modern method of inquiry, the use of analogy in discovery. He already knew that an object traveling with uniform speed would pass equal distances in equal time intervals. From that knowledge and a Pythagorean belief in the powers of pattern and order in nature's universe, it is easy to argue—though extraordinarily insightful—that an object undergoing free fall, which is supposedly uniformly accelerated motion, would gain equal increments of speed in equal time intervals. In other words, if the speed of the freely falling object at the end of the first second was, say, 32 feet per second, then the speed at the end of each of the succeeding seconds would increase in increments of 32 feet per second—from 32 it would increase to 64, then 96, etc.

Salviati argues that, starting from rest, a freely falling body acquires equal increments of speed in equal intervals of time. But this is simply part of the definition of uniform acceleration, the kind of acceleration a freely falling body undergoes.

What follows is a marvelous argument from Sagredo, very suggestive of a Zeno-like paradox. Sagredo argues that there is some strange contradiction in thinking that starting from rest a body gains speed in proportion to time. Measure time backward in intervals between pulses. Suppose that at the end of the fourth beat the body has a speed of two units. Then at the end of the second beat it would have been traveling one unit. Since time is divisible without limit, it follows that the earlier speeds are less than the later. Continue backward in time intervals approaching the instant when the body first started, and we find that the body must have been moving so slowly that it could never have started.

"We must infer that," he says, "as the instant of starting is more and more nearly approached, the body moves so slowly that, if it kept on moving at this rate, it would not traverse a mile in an hour, or a day, or in a year or in a thousand years; indeed, it would not traverse a span in an even greater time; a phenomenon which baffles the imagination, while our senses show us that a heavy falling body suddenly acquires great speed." But Salviati tells us that he, too, had been puzzled by the notion that speed was proportional to time. He tells us that he investigated the hypothesis by performing an experiment. He placed a heavy ball on material that flexed under the weight and marked the amount of flex. Next, he raised the ball to a certain height and dropped it onto the flexible material and again measured the flex. He repeated the experiment, each time raising the ball higher, and noted that the amount of flex increased as the height increased. From this he concluded that the velocity of the ball must have been increasing with height.

In another experiment, he took a piece of wood, cut into the wood an extremely straight groove, smoothed the groove and lined it with parchment. The board was placed in an inclined position. A very smooth and polished round bronze ball was rolled down the inclined plane. The time of descent was recorded. This was repeated many times for accuracy by taking averages "in order to measure the time with an accuracy such that the deviation between two observations never exceeded one-tenth of a pulse-beat. When he rolled the ball from a quarter of the height, he found that the time of descent was one-half that of rolling it down the full length of the groove. He repeated the experiment one hundred times and always found that the distances traversed were to each other as the squares of the times. Again, he repeated the experiments with the board at various angles of incline and found the same rule: the distances traversed were to each other as the squares of the times."

Galileo knew that an object moving along a frictionless horizontal plane would continue to move indefinitely with a uniform motion. In day four he introduced the following remarkable theorem: "The spaces described by a body falling from rest with a uniformly accelerated motion are to each other as the squares of the time intervals employed in traversing these distances."

In other words, if the falling body moves x feet in the first second, it will move a total of $4x$ feet by the end of the next second, $9x$ feet by the end of the third second, and n^2x feet by the end of the n-th second. There is another way of looking at this: The distances traversed in each second will be in the same ratio as the series of odd numbers $1, 3, 5, 7, \ldots$, which means that if the falling body travels x feet in the first interval, then it will travel $3x$ feet in the second, $5x$ feet in the third, etc. Pythagoras would have been thrilled with this discovery. It plainly confirms that numbers are clues to understanding the nature of the physical world, and that the universe is ordered and explainable.

The explanation: When a body falls from rest with uniform acceleration it moves n times as fast in the *n-th* second as in the first second—twice as fast after two seconds as in the first second, three times as fast after three seconds as in the first second, etc. Also, the body will fall four times as far in twice the time. Notice, he was not saying how far it will fall, just that if it falls x feet in t seconds it will fall $4x$ feet in $2t$ seconds. Another way of saying this is that it will fall three times as far in the second second than the first.

This is precisely what William Heytesbury and the Merton College mathematicians were saying 250 years earlier. Surely, Galileo must have been aware of the acceleration theorem argued by the Merton mathematicians, which was proven by Nicole Oresme in the mid-fourteenth century. Yet he does not mention this in his writing. Galileo's contribution is his brilliance in experimentation. Timing instruments were too crude to detect and measure speeds of free fall to the second. Those timings could be off by as much as fifty percent.

The inclined plane could be made shallow enough to record speeds, times, and distances. So he could take a plane that slows balls down to speeds at which the ball travels two feet in the first second. Assuming that the acceleration is uniform, that would mean that after two seconds the ball would be traveling at eight feet per second; after three seconds it would be traveling at twelve feet per second, etc.

Now, here's the critical move. The speed is increasing, but it is increasing at a constant rate; so if the ball started rolling from rest, at the end of any time interval its speed must be the average speed over the interval multiplied by the time elapsed. Galileo would have observed that the distance traveled by the end of two seconds would be (4 feet per second) $\times$ (2 seconds) = 8 feet.

At this point, Galileo must have seen a magnificent law: The ball will always roll four times as far when the time interval is doubled. This is reasonable, when one considers that uniform acceleration means a constant increase in speed. This is what actually happens. Since the speed is constantly increasing, the average speed over two seconds must be double that of the first second. So Galileo considered the time interval t. He knew that the average speed at the end of the time interval 2*t* must be double the speed at the end of interval *t*. Hence the distance traveled by the end of 2*t* seconds may be computed by realizing that the ball must be moving at twice the average speed for twice the time. Hence, the ball rolls four times as far when the time interval doubles. This same reasoning shows that the respective distances traveled in 1, 2, 3, 4, . . . seconds are the squares 1, 4, 9, 16, ...

The Pythagoreans had mystically identified numbers with nature, hitting on some very important relationships, 2,100 years before Galileo experimented with falling objects. But Galileo was discovering that these mathematical patterns were snugly identified with nature and that they could, in turn, be generalized and used to subordinate nature by predicting what will happen. Thus—just as the Pythagoreans noticed that squares behave in a snug relationship to each other when on the sides of a right triangle—Galileo and his contemporaries could see that squares might arise from experiments with falling objects and conclude that, perhaps, areas and the geometry of space might have something to do with motion.

· 6 ·

Dance of the Planets

The world, and especially the church, had accepted the great astronomer Claudius Ptolemy's earth-centered model of the universe since the second century. Ptolemy's theory was supported by centuries of astronomical observations and the simple geometry of circles. It confirmed the biblical passages of Joshua and was consistent with Genesis, so the church was happy to back that model with strong support. But as new observations and increasing knowledge mounted, more and more complicated adjustments and amendments became necessary to keep the theory in line with raw data.

As new heavenly phenomena were spotted, more intricate explanations were added. At first a few new cycles were needed to fit new observations of planetary motion, observations from earth (which was thought to be stationary but was actually moving) that gave the impression that at times the planets take little circular paths before continuing on their orbits. Then, when Mars was discovered to periodically display retrograde motion, epi-epicycles were added. This retrograde motion should have been considered peculiar in a perfect universe of circular motion, especially if the earth is at the center, because then the planets and sun should always be moving in one direction. But retrograde motion was clearly observed. Mars would slow down, come to a halt, reverse direction, come to another halt, and reverse direction once again before continuing. After 1,300 years of added complexities, a small group of astronomers felt it was time to rethink the theory.

NICOLAUS COPERNICUS WAS born on February 19, 1473, in Torun, Poland, a picturesque medieval walled town on the Vistula River. He died in 1543, the year his epic work *De Revolutionibus Orbium Coelestium* (*On the Revolutions of the Heavenly Orbs*) was published. In the century before, an epic work such as *De Revolutionibus* would have hardly been seen by anyone more than a hundred miles from Krakow, but printing contributed to the first phase—perhaps the dusty dirt roads—of an early information highway. Though few astronomers were capable of reading Copernicus's thirteen-volume book, it was now accessible to experts, and its most radical point was clear: The sun is the center of the solar system, and the earth is just one planet like any of the others that revolve in space around the sun. It was not a new concept, but Copernicus gave it new life through purely mathematical support. At its early stages, his theory was taken as an interesting fiction. Like all fiction, it was not taken seriously and therefore not considered blasphemous. Besides, what could the church do? By the time he was taken seriously, he was dead.

Copernicus's model was simple. Put the sun at the center of the universe, let the earth and planets orbit in circles around the sun, and let mathematics take over. Fewer assumptions were required to explain the movements, and the entire theory was mathematically simpler than Ptolemy's. But it would take more than a few assumptions and mathematical simplicity to convince those who grew up believing that the earth was immobile.

Tycho Brahe did not fully believe Copernicus's model; he sensed that something did not fit observable facts. As a seventeen-year-old at the University of Leipzig, he observed a meeting of Saturn and Jupiter, which according to both Ptolemy and Copernicus should have occurred on a different date. The Ptolemaic model gave a much wilder prediction than the Copernican, but it shook Brahe's confidence in both. He developed his own theory of a sun and moon orbiting a stationary earth and the other planets revolving around the sun.

Brahe proudly groomed his handlebar mustache to extend well beyond his cheeks and entirely cover his mouth. But it did not detract from the prosthetic copper nose bridge that replaced his real nose bridge, which was mulilated in a duel when he was a young student at the university in Rostock on the Baltic Sea. Tycho was only twenty on a December night in 1566 when he met up with Manderup Parsbjerg, a fellow Dane of Rostock, at a dance. Tycho and Parsbjerg began to drink heavily and argue over a young lady, when, with predictable sixteenth-century high regard for honor, the argument led the two into dark woods behind the university for a duel that cost Tycho his nose.

At the relatively young age of twenty-six, Tycho began

constructing an observatory at the Herrevad Abbey, near Copenhagen. The telescope had not yet been invented, so his observations were made with the naked eye. On the evening of November 11, 1572, after emerging from his alchemy laboratory, Tycho spotted a brilliant white object directly overhead, a bright new star in the constellation Cassiopeia, one that had not been seen before. Even with the instruments available to him he could see that it did not shift position with respect to the background. For the remainder of the year, the star could be seen to change from white to red, and then to gray. He concluded that it was very far away, much farther than the moon, whose shift of position with the background could be measured. But more interesting to him was the newness of the star. If the celestial world was perfect and unchanging, as Aristotle had professed, then how could a new star appear?

Was it a star? He built a compass device to accurately measure the star's latitude and longitude to detect any movement. Any perceptible motion would indicate that it would be only as far away as the moon and not a star. If it were not a star, that would not refute Aristotle's notion of a perfect unchanging heaven, because things as close as the moon would by nature be imperfect, corrupt and changing. But if the object that Tycho saw was a star, it would challenge the purity of heavenly stillness. There was not the slightest movement, so it must have been a star.

We now know that Tycho was observing a supernova—either the birth or death of a star, but in either case a grand explosion. New stars had been noticed long before Tycho's discovery. When Hipparchus sighted one back in the second century BCE, he compiled the first star map against which future stars could be logged. In 1054 an astonishing new star in the constellation Taurus was brighter than Venus. It could even be seen in daylight.

Tycho's new star, after suddenly appearing northwest of the constellation Cassiopeia and brighter than the brightest planet, soon disappeared, never to be seen again. There had been many star changes in the past, but none aroused much curiosity about the immutability of heaven. The remarkable observation coming from Denmark's greatest astronomer could have been enough to loosen centuries of entrenched belief in Aristotle's strict doctrine of an unchanging heaven, but it didn't.

Five years after Tycho discovered his supernova, another rare event took place. He observed a bright star with a red tail. After observing its motion and computing that its position was far beyond the moon—more than four times the distance of the moon—he could conclude that the red-tailed object was indeed a bona fide comet, corrupting Aristotle's perfect heaven.

For the next twenty years, night after night—even in the dead of winter, when he tried to keep warm under his heavy woolen hooded robes—he spent his time systematically building and improving his accurate instruments, inventing new ones, and cataloging the positions of all known astronomical objects with astonishing accuracy.

By the turn of the seventeenth century, when he was appointed Imperial Mathematician to the Holy Roman Emperor, Rudolph II, Tycho had moved many of his instruments to Prague, the capital of the Holy Roman Empire. He intended that this work should prove the truth of his cosmological model, in which the earth (with the moon in orbit around it) was at rest in the center of the universe and the sun went around the earth (with all other planets being in orbit about the sun and thus carried around with it).

It now appears astounding that the Copernican picture of the universe was at that time still a philosophical one. Copernicus's arguments, suggested by geometrical simplicity, had been dismissed as an abstract diagram that bore no relation to reality. But in 1610, almost seventy years after Copernicus blasphemously suggested that the earth was not the center of the world, Galileo used one of the first astronomical instruments to observe the motion of the planets.

Galileo gave the first serious but informal proof of the Copernican theory, for which he was imprisoned by the Inquisition. One now reads with amazement:

> There was published some years since in Rome a salutiferous edict, which for obviation the dangerous scandals of the present age, imposed a scasonable silence upon the Pythagorean opinion of the mobility of the earth. There want not such as unadvisedly affirm that that decree was not the production of a sober scrutiny but of an ill-informed passion: and one may hear some mutter that consultors altogether ignorant of astronomical observations ought not to clip the wings of speculative wits with rash prohibition.

In early seventeenth-century Bohemia, complex religious and constitutional tensions were brewing. All that was needed to trigger the war that would spread throughout Europe and last for thirty years was a gang of Protestants to throw two Catholic governors from a Prague castle window. Until then, Prague was the great center of European science and alchemy, an attraction for many eminent scientists.

There, on the eastern side of the Vltava, Johannes Kepler, now the Imperial Mathematician, worked on his models of the solar system. After years of gruesome calculations, based on Tycho Brahe's recordings and observations, Kepler would formulate three laws that govern planetary motion. It is astounding that Kepler's three laws could be deduced purely from observations of the sky. He did not know why it should happen, but he must have known that his work implied a harmony between the world of observable facts and that of purely rational mathematics. He had exposed a miracle; an overwhelming amount of facts fit into a few brief verifiable statements revealing glorious relationships between space and time.

Ideas of perfect symmetry come into the human mind more naturally than those of asymmetry. It was no wonder that Aristotle, just as the Pythagoreans before him, imagined circular motion to be the only perfect nonlinear motion. So when he thought of heavenly bodies, he could only imagine them moving in circles. Once an idea like that enters the human mind, it embeds itself so deeply it becomes difficult to uproot in order to make room for others. In Kepler's youth, most people believed that the planets were carried around in circles by angels. The angels were not the problem, the circles were. At the same time, from a young age, he strongly believed in Copernicus's theory that the earth revolved around the sun and persisted in questioning the connection between the orbital distances of the planets and the times they took to orbit the sun. He knew that the farther the planet, the slower it appeared to move.

He had an appealing, yet groundless, idea. Start with a circle, inscribe an equilateral triangle and rotate the triangle in the circle. These rotating triangles will envelope a smaller circle of radius of half the size of the original circle. "Perhaps," he suggested, "these circles correspond to the orbits of planets." What happens if squares are inscribed? The radius of the smaller circle is $1/\sqrt{2}$ smaller. He then tried hexagons and other polygons, but in the end, he had to give up the idea.

Of course, such an idea is fanciful. Why should planets behave like polygons in circles? Such a question must have entered his thought at fleeting moments, but his first concern was *how* planets move rather than *why* they do. Any answer would have been strongly influenced by the Galilean notion that the world follows mathematical order. Many such ideas were based on searches for pattern or regularity. One of Kepler's better ideas was to use the symmetries of the five regular Platonic solids—those solid figures built from surfaces whose faces are all identical.

These could be traced back to the Pythagoreans who may have abstracted the idea from crystals of pyrite, a sulfur mineral natural to the hills of Sicily. In Plato's *Timaeus*, these Pythagorean solids are used to represent fire, air, earth, and water, with the dodecahedron reserved as the image of the entire universe. The great mathematician of Plato's academy, Theaetetus, constructed the dodecahedron; Euclid had proven that there are only five such regular solids. Perhaps the five solids could nest in each other in such a way that they define spheres with the property that their radii correspond to the radii of the six planetary orbits. That would have been wonderful, for it would have also explained why there were only six planets.

In 1595 Kepler was about to abandon his idea of inscribing polygons in circles when he dreamt up a new idea—an idea so splendid that he feared he had unlocked a divine secret. He hesitated to publish it. But he wrote about modeling the universe by alternately circumscribing spheres and Platonic solids. He represented Mercury's orbit as a spherical shell (whose thickness represented the difference between the planet's minimum and maximum distance to the sun) and enveloped it with an octahedron. This octahedron, in turn, was enveloped in another sphere. This new sphere represented the orbit of Venus. Continuing, he enclosed the sphere of Venus in an icosahedron. This he enclosed in the sphere of Earth; then a dodecahedron with a sphere to represent the orbit of Mars; a tetrahedron with a sphere to represent Jupiter; and finally a cube with its sphere to represent the orbit of Saturn.

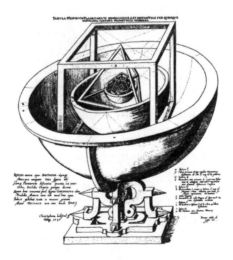

It was a glorious idea. He was very proud, enthusiastic, and hopeful. He claimed, "The intense pleasure I have received from this discovery can never be told in words. I regretted no more the time wasted; I tired of no labour; I shunned no toil of reckoning, days and nights spent in calculations, until I could see whether my hypothesis would agree with the orbits of Copernicus, or whether my joy was to vanish into air."

The real question was how closely would the orbital distances agree with Tycho Brahe's observed data? They turned out not to agree very well, so the model dissolved into a purely fictitious image of the universe. He must have played with different orderings of solids to find this best possible spacing. If it had worked, it would have fallen into the medieval Aristotelian trap of suggesting symmetry as a prime cause of movement. What would it have said about Uranus after it was discovered in 1781? Of this, Hermann Weyl said in his famous book *Symmetry*, "We still share his belief in a mathematical harmony of the universe. It has withstood the test of ever widening experience. But we no longer seek this harmony in static forms like the regular solids, but in dynamic laws."

Kepler was assuming that planets orbit in circles with the sun off center. The circular orbits presented a distinct problem. Just as in the old Ptolemaic system, the orbits would appear to once in a while do a little dance—a retrograde, back-and-forth path. This retrograde motion was not eliminated by the new Copernican idea, and predictions of where a planet should be were not good enough. Kepler tried to fit Tycho's data to all sorts of circular orbits, adjusting his models with circles within circles and epicycles along epicycles, groping for some appearance of a theory or law. Frustrated in his attempts, he often wondered if Tycho's data was wrong, but quickly banished such thoughts. He knew that the planets move in nonuniform motion faster near the off-centered sun, and slower away from the sun. "What if," he thought, "the area bound by the circular orbit is divided into equal areas meeting at the sun?" It was a lark of a thought; but hold on. Perhaps the speed of Mars varies in such a way that the planet moves across areas in equal times. . . .

It was a wild thought, one that must have come from concentrated study of the data. For a short time, this seemed to be right, but there were other problems. His rejoicing was premature.

"While thus triumphing over Mars," he wrote, "and preparing for him, as for one already vanquished, tabular prisons and equated eccentric fetters, it is buzzed here and there that the victory is vain, and that the war is raging anew as violently as before. For the enemy left at home a despised captive has burst all the chains of the equations, and broken forth from the prisons of the tables."

It is remarkable that, until that moment, nobody had considered other curves. Surely, ellipses and other conic sections had been thoroughly studied by Apollonius in the third century BCE. But Apollonius and his conic sections were not well known in Kepler's time. Besides, they would have been considered too impure for celestial orbits. The circle was the most perfect of all curves, and if the heavens were to be turned by God or the angels they must be made from circles.

The title, Imperial Mathematician, didn't mean much, for Rudolph II, Tycho's patron, did not extend the same funding favors to Kepler as he had to Tycho. So Kepler spent his time brilliantly pursuing secrets of the universe, often working through the night without food or drink until daylight, when he would drift to sleep on a dusty sofa in the observatory. During those long nights studying Tycho's notes and data, laboring over massive calculations, he uncovered two clues to the mysteries of the heavens. We can imagine him half-asleep just at the break of dawn on one fine day quietly mumbling, "It looks as though the orbit of any planet is an ellipse." Then, with increasing adrenalin-kicking excitement, saying more loudly, "Yes! The orbit of any planet is an ellipse with the sun at one focus!"

He tried all sorts of oval orbits before thinking of the ellipse. The ellipse—how simple! In hindsight, we may think it simple, but the ellipse has a complexity that the circle does not. A circle has only one center. An ellipse has two *foci*. It is a conic section, a curve that results when a cone is sliced.

We can imagine the thoughts that followed. If the sun is at a focus of an elliptical orbit, then the earth and other planets must be eccentrically moving in imperfect orbits. Why should the sun favor one focus over the other? The motion is not uniform. The planets speed up and slow down. With the sun at one focus it is entirely possible that the eccentricity in geometry is linked to an eccentricity of motion. If the orbit is not circular, the distance from the sun to any one planet is not constant. So perhaps the planet speeds up near the sun and slows down as it moves farther from the sun. Incubating such thoughts, Kepler, again tired from another night's calculations, suddenly had another idea: *areas.* Using his hypothesis that the sun is at one focal point of an ellipse and Mars is in elliptical orbit around the sun, he drew two lines, one from the sun to Mars, another from the sun to where Mars was thirty days earlier. He found that the area proportion of the ellipse swept out by Mars in those thirty days to the full ellipse was slightly larger than 12 to 1. He examined the data again for the proportion of area swept out by Mars in sixty days and found it to be a bit larger than 6 to 1; in ninety days it was approximately 4 to 1; in 120 days, approximately 3 to 1; in 150 days, about 2.5 to 1; in 180 days, 2 to 1. In other words, in any thirty-day interval, the planet would sweep out an area approximately one-twelfth of the full area of the ellipse, no matter where the planet was. Excitedly, he formed the hypothesis that became his second law of planetary motion: *The line joining a planet to the sun sweeps out equal areas in equal times.*

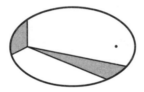

How does one stumble on the idea of governing the speed by the areas swept? It is such a remarkable thought, that one wonders if it is a stroke of genius or an accident of groping. The idea was revolutionary, for it marked a new sense of what governs motion. It was natural to think that the position of an object in motion would follow some geometric curve. But since the thirteenth century, when the Merton College mathematicians thought of the acceleration theorem, speed had been thought of as being controlled by the immediate events and properties of the object's locality. The moment Kepler made a connection between speed and swept-out area, speed was seen as being controlled by far more global causes. The divine secret of the movement of the planets was out of the box.

Kepler must have been extremely happy with his two laws, but they were not enough to convince die-hard skeptics of a heliocentric universe. It would take another ten years for him to uncover the clinching law. He considered the mean distance of Mars from the sun. It is 1.53 times that of the earth's mean distance. Its year is 1.88 earth years. Now 1.53^3 is very close to 1.88^2, a difference of less than five-hundredths. What about Venus? Jupiter? In each case, *the square of the period of a planet is proportional to the cube of its mean distance from the sun.* This is Kepler's third law. It completes the group of laws that will later become necessary for the clincher—the cause. But that cause will have to wait more than another half-century for Newton.

THE SEVENTEENTH CENTURY was different than the sixteenth. Through experiment and observation entwined with mathematics, physics would discover not only earthly phenomena but also universal marvels, and would grow beyond what anyone could have foreseen in the previous century. The seventeenth century belonged to Galileo, Newton, and many other clever experimenters and inventors who were supported by printing and the deliberate sharing of knowledge throughout Europe. Scientists no longer had to work alone, contemplating the ancient works of Aristotle or the liturgical dogma of the church. They had cafés, clubs, and meetings of scientific societies, the seventeenth-century equivalents of our Internet blogs and chat rooms. Of course, there were people who worked alone, not communicating their great ideas to others—monks who never left their monasteries, counts who hardly left their castles in Bohemia. Leonardo da Vinci secretly studied the human body by dissecting dead bodies. He also studied the movements of the earth about the sun. On paper, he designed airplanes, submarines, and parachutes. But he did not publish those ideas, and his brilliant sketches and ideas remained unknown until the twentieth century, when his notebooks were discovered.

Francis Bacon and René Descartes were in the vanguard of this new era of natural philosophy. They questioned the old methods of acquiring knowledge, they believed that there could be no reliable way of knowing nature with certainty. Medieval methods were foolhardy, they would say. We cannot find truth about nature by postulating something and deducing further truth. Truth about nature can come by rational thought only after lengthy investigation of nature itself and experimentation. Bacon gave a recipe for investigating nature in his *The New Organon*, which told us that to know the truth in science we must proceed from the particular to the general. By this he meant that we must start by observing many instances of a single phenomenon to isolate the core of the truth. To know that the tides are caused by the moon's gravitational pull, one must observe the tide at many times—low, high, and in between. Of course, in Bacon's time there were mathematical models of the tides and good reasons to believe that gravitational attraction to the moon was responsible. But truth is only as good as its mathematical model; observations and measurements are needed to convince the skeptic that the model is tightly reflecting reality.

Descartes was a mathematician who thought of nature as a dichotomy—mind, spirit, and consciousness on the one hand,

and the objective substance of everything outside the mind on the other. He wrote that it is possible to understand "the forces and action of fire, water, air, the stars and heavens, and all other bodies that surround us as distinctly as we understand the mechanical arts of our craftsmen. . . ." And that "we can use these forces in the same way for all purposes for which they are appropriate, and so make ourselves the masters and possessors of nature."

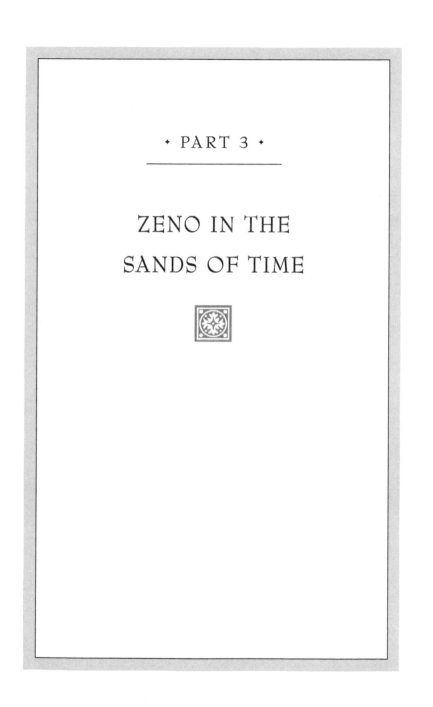

· PART 3 ·

ZENO IN THE
SANDS OF TIME

· 7 ·

A Step Back for Time

There are things we see and things we don't. We don't see the growth of a plant from second to second. If we see a plant cell divide, we wonder how one nucleus became two. If we see a nucleus become two, we look for the moment when an individual chromosome became a double chromosome. Still, in the mitosis of time, we look for the moment when the nucleic acids and proteins make their first move in the replication process. No matter how finely we divide time, we always find some discontinuity in the plant's development. It always comes down to the *one* becoming *two*. It all comes down to the end of the first paragraph of Genesis, when God divides "light from darkness," and those ancient questions first posed by Zeno and Parmenides. Continuity is a tool—and only a tool—to help us around the *how* in Zeno's arrow paradox. But Zeno has other queries in his quiver. To an atom sitting in one of his arrows, its neighbors seem a universe away. As the arrow rigidly moves, that atom will never catch up with its neighbor, which, presumably, has moved an equal distance. To the human observer, the time it takes one atom to move to the position its neighbor once had is immeasurably short. Yet, from the atom's point of view, it has moved an enormous distance. Even the most sensitive instrument may never observe such a minute shift in position. And if it could, then what could be said about half that shift, a quarter, an eighth, etc.? What clock could measure such shifts?

Zeno must have understood that time is entwined with the problem of continuous movement, and that space was messily mixed up with not only time, but with observation—which involves the whole question of position relative to a stationary object. How can one measure position, speed, or change in direction without referring to something stationary, a point of reference?

We get some sense of continuity from our direct experience with time and space, yet our modern conception of continuity transcends any familiarity with the real world. Ancient Greek mathematics had no concept of a continuous algebraic variable and no definition of an arithmetical continuum, the kind we now think of when we think of the real-number line. It took 2,500 years of work to get from the intuitive feeling for continuity of Zeno's era to the precise logical definitions of the late nineteenth century given by the mathematicians Augustin Louis Cauchy, Karl Weierstrass, and Richard Dedekind; in the end, we are left with abstractions projected far from the sensual world into subtle notions of the infinite, the infinitesimal, and fields of infinite convergent series. Continuity—once an exclusively visual impression of reality—has been amended to include a conception framed by the consistency of logical thought.

Though irrational numbers were treated geometrically in Euclid's *Elements,* they were not accepted as numbers before Newton's lifetime, when they became the model for applications of continuous motion and kinematics. Before the midnineteenth century it was known that there are holes in the rational-number line, though it would have been a huge surprise to find out just how frequently those holes appeared along the real-number line, yet the rational numbers furnished the physicist and engineer with rational approximations to any degree of accuracy so they could make predictions in the real world.

The set of real numbers (those numbers that can be expressed as a—possibly infinite—decimal expansion) may be represented on the number line by the set of points extending infinitely in two directions from zero. The numbers themselves may be thought of as representing distances from zero according to some scale—negative numbers to the left of zero, positive to the right. Such a line can only be imagined, but an illustration may help. Let's examine a short interval within the infinite number line, say, just the real numbers from 0 to 1. Draw a line interval whose length is one unit. Measuring from left to right, the first point on the left edge of the interval represents the number 0; a point at distance d units (less than 1 unit) from the left edge represents the number d. In the illustration on the next page, the point we are calling 1/4 is one quarter of a unit from 0; the point we are calling $\pi/4$ measures $\pi/4$ units from 0. A point P is called a *rational point on the interval* if P is a point whose distance from the left end is a rational number of units. *Irrational points* are defined in a similar way.

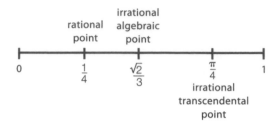

The set of rational numbers is a relatively small subset of real numbers—those real numbers that can be expressed as fractions (or equivalently, those whose decimal expansion is finite or that eventually form strings of a repeating pattern).

For the dichotomy and Achilles paradoxes both space and time are examined only at rational distances from the starting position. The dichotomy case questions what happens at distances $1/2^n$ units; in the Achilles case, if the speeds of the racers are rational numbers, and if the tortoise's head start is a rational number, then each point under consideration in Zeno's argument is a rational distance from the starting line.

One big question behind Zeno's motion paradoxes is this: If the Zeno arrow is moving through a point represented spatially by a real number of the number line, then how does it get to the "next" point along its trajectory when there is no next point on the number line? A notion of next point is meaningless in the geometry of the real- (or even the rational-) number line. For example, take π; its decimal representation is 3.141592654. . . . The trail of digits indicated by the ellipsis of dots is infinite. So what is the next number after π? Or, take the rational number 1/2, which, in decimal notation, is 0.5. What is the next rational number? It cannot be 0.51, nor 0.501, nor 0.5001, nor any number starting with a decimal expansion of 0.5 and ending with some long string of 0's with a 1 at the end, for such a number would be further from 0.5 than one gotten by slipping in another 0 before the final 1. So if the tip of the arrow has traveled, say, 1/2 its anticipated distance, where does it go next?

Late in the summer of 1872, Georg Cantor, the Extraordinary Professor of Mathematics of Halle, Germany, had a shocking revelation that there are far "more" irrational points on the number line than rational points. If rational numbers were the only numbers represented, then the number line would have holes EVERYWHERE! Between *any* two rational points there would be not just one hole but also an infinity of holes, so the number line would be far from being a dependably continuous line.

What is said for space may be said for time. What is time? Something happens and something follows. One event follows another in a sequence that must be comprehended somehow. The arrow moves from one place to another. *Before*, it was there; *now* it's here; *later* it will be there; and far in the *future* it will reach its destination. These are the raw materials of time.

Primitive humans did not have a concept of "five minutes," though they must have had the notion of time passing as the sun continuously passed from rising to setting. Precision would only come as a result of experience and need. Fish bite more frequently in the morning; caribou graze in open plains by day; and not much can be done at night.

> And God said, Let there be light: and there was light.
> And God saw the light and it was good: and God divided the light from darkness. And God called the light

Day and the darkness he called Night. And there was
evening and there was morning, one day.

According to the Bible, we were simply given a division of one
day, which repeats. From it, we have created a convenient
scheme for recordkeeping. We invented the afternoon. Eventually, by the time of the ancient Babylonian and Egyptian civilizations, when human affairs became so complex that time
required more precision, the day was broken into twenty-four-
hours. It was a clock of heaven. The nighttime hours were divided by twelve groups of stars that appeared in the sky; the
daylight hours were divided into twelve, to match.

For years, the sundial and the water clock—a tank filled
with water leaking at a nearly constant rate, with indicators
mechanically controlled by floating bobs connected to levers
marking time—governed our days without hints of an hour or
minute. The technical difficulty of using the sun to break down
time into shorter intervals was linked to the complex vernal
shifts between the seasons.

Both the ancient Egyptians and the Romans had water
clocks. By the fourth century BCE, they had the idea to divide
the day into two parts, our A.M. and P.M. (*ante* and *post merid-
iem*). Later, the day was divided into quarters—early morning
and forenoon, afternoon and evening.

By the first century CE, Romans were getting more sophisticated. Daylight hours were treated differently than nighttime
hours. At the height of winter, when the sun shone for a bit less
than nine hours (by our meaning of *hour*), the Romans would
still break the daylight hours into twelve forty-five-minute seg-
ments (by our meaning of *minute*). In the summer, this would

be reversed. Their water clocks should have been reset each day, but even the best clocks could not be calibrated so finely. So once a month, an official timekeeper would reset all the clocks of Rome.

Later, Christian monks and Muslim clerics needed a scheme to call others to prayers. Monks in European monasteries devised mechanical contraptions (alarm clocks) driven by weights that struck bells to awaken a bell ringer who would ring the larger bells atop towers. The mechanical clocks were set to markings prescribed by church canonical hours—sunrise matins, noon none (the ninth hour counting from sunrise), evening compline, and nighttime vespers—not to equal divisions of day or night. Our English word *clock* is derived from the German *glocke*, which means *bell*.

Slicing the time of day into minutes and seconds came later, when punctuality in commerce became critical for organized appointments, shipping, and local travel. Such thin slicing required more accuracy than the sundial or water clock could give. It required a mechanical device that could count equally spaced moments without the help of intermittent sunlight or continuously flowing materials such as water, sand, or burning oil.

Some of the greatest inventions supporting human progress never get the credit they deserve. We talk of the wheel, the bow and arrow, the lever, the hammer, the steam engine, the screw, the fitted sheet, etc. These deserve high praise. But, aside from the screw, these inventions are accidents of observation. How many rolling logs or pomegranates does it take to notice the idea behind the wheel? How many whipping branches does it take to notice a way to utilize the elasticity of bending wood?

How many observations of logs resting on stones does it take for children to discover the seesaw or adults to sense the power of a lever? Many of these devices were discovered, not invented. They were models of principles that nature left lying around in plain sight; almost anyone could have picked one up. Civilization would not have advanced much beyond the Paleolithic age without some of these inventions, but there is one invention whose acclaim is long overdue. Horologists know it as the *escapement*.

The earliest clocks used water, sand, or oil, but their need for perpetual maintenance limited their continuity and accuracy. The problem with time is that it is both continuous and regular. What in the world—other than time—has those features? Even the human pulse, which Galileo is reputed to have used as a measure of regularity, often changes after short intervals. At first thought, it may seem easy to keep a gear moving at a steady, continuous rate. Think further and it will become clear that this may be one of the world's most difficult problems. For the past 900 years, most mechanical clocks have used oscillatory motion, with the same general principles guiding inventions of horology.

In principle, there are several different basic oscillatory generators. Take the case of a stationary vertical spring with a weight attached. Extend the spring by pulling down on the weight and the spring will oscillate up and down, losing energy to air resistance and heat through the molecular forces of expansion and contraction. When the spring is stretched or compressed it exerts a restoring force proportional to the length of extension or compression—Hooke's Law. Pull a hanging pendulum bob to one side and it will swing past the vertical and back to its initial height before returning to repeat the cycle. It too will lose energy to air resistance and friction at its pivot point. Its amplitude will gradually diminish, but as Galileo noticed, the period of oscillation will not depend on that amplitude.

The pendulum can do two things: it can count (in oscillations) and it can—with a bit of help—maintain its swing. But these are two different functions. To do both we need the es capement.

It's possible that the Chinese invented the earliest escapement in the eleventh century. Its inventor, Su Sung, built an enormous Rube Goldberg contraption several stories high that used a turning water wheel of buckets that scooped and spilled water to tilt levers that alternately caught and released sprockets attached to the wheel. Though it may be a stretch to suggest that this Chinese clock fully used the principle of the escapement, it did use a scoop and tilt mechanism to regulate time.

As early as 1286 Saint Paul's Cathedral in London had a clock with a clock steward named Bartholomew whose pay included a loaf of bread and some beer. About fifty years later, Walter Lorgoner made an improvement by giving the clock a turning angel, for which he was paid six pounds sterling, even though he had to bear the cost of "iron, brass and all manner of things for carrying out the said work." If a weight is attached to a rope wrapped around a cylinder attached to a gear mechanism that turns a clock dial, the weight will simply fall and pull the rope, which turns the cylinder, which turns the clock dial. The clock dial will quickly spin until either the weight reaches the ground or the rope fully unwinds from the cylinder. The turning dial will not have measured time, but rather have measured the time it takes for the weight to fall or the rope to unwind.

The problem is how to keep time moving in a regular fashion. The solution is the escapement. The first mechanical one aside from Su Sung's interesting contraption was the *verge escapement*. It remained in its primitive twelfth-century form for 400 years. It works like this: First, there is an escape wheel called the *crown* because it looks very much like a royal crown with triangular teeth curved in the direction of rotation. (See figure on p. 107.) The axle of the wheel is horizontal. Two weights counterbalance the crossbar (*foliot*). The crown is being driven by the motor force, which is likely a weight hanging from a rope coiled around the axle. The foliot and verge are manually set in rotating motion causing one pallet, say the top one, to make contact with the highest tooth of the crown. The foliot and verge continue to rotate until the top pallet clears the top tooth of the crown and the lower pallet (at a right angle to the upper pallet) comes in contact with the lowest tooth of the crown, forcing the verge and foliot to stop. The impulse resulting from the quick stop gives enough of a shove to the pallet to restore any energy lost from its last rotation, causing it to rotate in the opposite direction. A complete rotation back and forth creates a unit of time, which ordinarily would be translated to some indicator that that unit of time has passed.

One problem with this marvelous contraption was that a verge clock requires a large rotation. Its clever unknown inventor probably saw it, too. The pallets need sufficient clearance to rotate in and out of the crown's teeth; in many cases the required rotation is more than twenty degrees of arc. The prob-

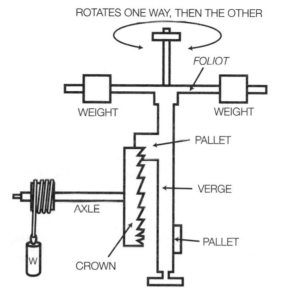

lem is that a verge rotates in a circular arc and such an arc does not follow the simplified formulas for period (the time it takes for the pendulum to make a complete swing). Any error grows with amplitude. A verge rotating in an arc of twenty degrees would lose more than a quarter-hour a day, thereby requiring adjustments. Moreover, periods of pendulum swings depend on pendulum length, which depends not only on latitude, but on climate. The idea was good, but the escapement needed to be improved to measure time usefully.

The best of many ingenious improvements was the *anchor escapement,* pictured on the following page. If the amplitude of the rotation could be reduced to two degrees, then only 6.6 seconds would be lost in twenty-four hours. The anchor escapement did not require very much of a rotation.

This small gadget, which can sometimes fit inside a seamstress's thimble, is responsible for a fair share of modern civilization. One of its functions is simple: to regulate and continue circular motion in mechanical clocks by alternately checking and releasing the teeth of a rotating gear, one tooth at a time. But it has a secondary function—to transfer some energy back to the pendulum or hairspring. Take the case of a spring: A small flywheel is forced by a wound spring to turn as the spring relaxes. The flywheel builds momentum to overshoot the relaxed state of the spring, and thereby rewind the spring and return to repeat the cycle. In this way the flywheel alternates between clockwise and counterclockwise rotation. Each change in direction alternately lifts each side of the escapement to release its grip on a turning gear, one tooth at a time. The ticktock sound of the mechanism is the escapement alternately catching and releasing a gear tooth.

Nobody knows for sure who invented the anchor escapement, either. Was it William Clement the London clockmaker, who in 1671 built at a cost of £40 a long pendulum anchor escapement clock for King's College, Cambridge, or Robert Hooke, the accomplished physicist who presented an anchor escapement clock to the Royal Society in 1666 after the Great Fire of London? Whoever it was must have understood that though time may seem—or even be—continuous, only discrete blips can measure it.

Later improvements—such as introducing a coiled spring, rather than a hanging weight, as the driving force—enabled clockmakers to miniaturize their works. But the basic idea remained: the transmission of energy from a source, such as a falling weight or an unwinding spring, to some oscillating motion that tracks the flow of time. In some timepieces a counter moves backward one position to move forward two. Yet in every case, in every clock, from Su Sung's Chinese water clock impulsively counting by oscillating buckets of water, to modern atomic clocks using the natural frequency of cesium-133—which oscillates at more than nine billion cycles per second—a discrete counting process measures time. Even the simplest modern watches rely on quartz crystals vibrating at more than a hundred thousand times per second.

The way time is measured is at odds with how we think about the nature of time. We think of time as moving smoothly in one direction, not two; why do we need a mechanism that reverses and repeats? In every case, we measure time by some form of stop-and-go mechanism. Zeno's arguments remind us that time might not be as continuous as it seems. Is there an elemental unit of time that cannot be split? Could time, like light, be composed of minuscule particles, quanta? The German genius of quantum mechanics, Werner Heisenberg, once suggested that the smallest unit of time is something in the neighborhood of 10^{-26} seconds. Modern physics can detect the difference in time intervals as close as one-trillionth of a second. Indeed, what we have been calling a clock turns out to be anything that has a countable uniform oscillation; if we can find some atomic beam vibrating with a period of 10^{-26} seconds, we would have the universal clock—if Heisenberg is right. But what if time is continuous?

Zeno conjured us into thinking that his arrow moved in discrete jumps, that Achilles was always catching up to where the tortoise once was, that we move across a room by considering half the distance we intend to cover. It took us a while to realize that another way of considering the paradox of motion is the paradox of measuring time.

· 8 ·

Descartes and the Magic of x and y

If you were the proverbial fly on the wall in Descartes's bedroom in La Flèche, in the south of France, in 1636, you might have seen Descartes lying in bed watching you. His most brilliant idea came to him while watching a fly crawl along a curved path, which he thought about describing in terms of its distance from the walls. A revolution in thought was in the making; mathematics would never be the same.

The German philosopher Daniel Lipstropius, a contemporary and biographer of Descartes, invented this fable. It later inflated into a more sweeping fiction of how Descartes, because of his poor health, would remain in bed late each morning meditating on how all of science could be made as certain as mathematics.

If the story were true, a fly would be responsible for one of history's most radical shifts in understanding mathematics and how it works. It would be responsible for an early marriage of algebra and geometry. If the fly traced a curved path in space, it would also have left a trail of arithmetical data, and Descartes would have understood that the geometry of the curve could be reconstructed from the arithmetical data—and, conversely, that the arithmetical data could be reconstructed from the geometry of the curve, and that geometry and arithmetic were simply different interpretations of the same mathematics.

There is substantial evidence that Descartes would lie in bed till late morning submerged in concentrated thought about his existence and natural surroundings. It was a habit that had lingered from his youth, when the boy was permitted to lie in bed to nurse his uncontrolled dry coughs, which seemed to fade by afternoon. Many such thoughts enveloped a belief that the physical world is fundamentally mechanical; everything in nature can be explained through the laws of mechanics; if the world is truly mechanical, then all theoretical physics should be expressible through a small number of general laws. His analytic geometry expressed the geometry of mechanics through algebraic equations and greatly helped to reconcile the observable facts of nature with a truly small number of principles and fundamental equations.

Descartes ingeniously used algebra to find the shortest distance from a point P to a curve. This was an ancient problem of conic sections addressed by Apollonius in the third century BCE. Descartes's imaginative method was to construct a generic circle centered at the point P. If the circle crosses the curve, it will normally do so at two places. But if it touches at only one point, the radial line from the circle's center P to the point of contact between circle and curve must be perpendicular to the curve. The radius of such a circle will give the smallest distance from the point P to the curve. Find the circle centered at the point P touching the curve at only one point (call it Q) by looking for a solution to the simultaneous equations of the curve and generic circle. But with that radial line from P to Q comes—free of charge—the tangent of the curve at the point of intersection Q. It is just the line perpendicular to the radial line. If the circle represented the orbit of a planet, then the tangent line at P is the unique direction in which the planet is moving when it is at position P.

With this new insight, Descartes set up the scaffolding for calculus, securing it for Newton and Leibniz's climb. He understood the importance of the tangent to a curve and developed a means of finding it by taking a circle that intersects the curve at two points and resizing the circle so the two points coincide. The tangent to the curve then becomes the tangent of the resized circle. This resizing process had hints of what was later to become one of calculus's great achievements.

Descartes didn't claim to be standing on the shoulders of giants, but he did owe credit to others who came before him. Menaechmus, in the fourth century BCE, discovered connections between conic sections and equations; early Greek geographers surely made free use of coordinates systems; Nicole Oresme, in 1361, worked with a system of latitudes and longitudes introducing early ideas of a coordinate system, complete with a horizontal line to represent time and a vertical line to represent speed; François Viète's creative notation relieved Descartes's unwieldy algebra considerably; and Fermat discovered the relationship between extreme values and horizontal tangents to curves.

Besides the seemingly miraculous notion that geometry and algebra are mirror images of each other, Descartes's coordinate geometry contributed two vital ingredients to mathematics: (1) the easy calculation of the distance between any two points using the Pythagorean theorem, and (2) the ability to represent straight lines and conic sections by equations and proportions.

A curve was no longer a static figure as it was for Greek geometers, who thought of classical conic sections—parabolas, ellipses, and hyperbolas—as curves formed by planes that cut cones. Instead, a curve began to be thought of as dynamically moving points determined by a rule (its equation), as an algebraic object with addresses (i.e., points) indicated by real numbers x and y. Those real numbers, the *coordinates*, were locked together in a co-ordered numeric relationship; one could not change without the permission of the other. This new geometry looked at curves as relations between variables; it was a very great advance, one that radically changed the tactics and manner of mathematics, one that made calculus possible, and one that changed forever how we think about Zeno's paradoxes of motion.

A typical sixteenth-century empirical observation would have shown the height of a projectile at various times as a table of values. But there was no rule for knowing heights at times when the projectile was not observed. Crude understanding of projectile motion may come from a table such as the one on the next page; yet a much more practical scheme comes from two other bookkeeping devices—the graph, which is helpful in giving an intuitive picture of how the numbers are climbing or descending, and the algebraic equation relating a height h to any time t. For example, the table below does not give the height at 2.5 seconds (the maximum), nor does it give it for non-integers, but the simple equation $h = -16t^2 + 80t + 1$ gives the height h at *any* time t.

Time in seconds t	Height in feet h
0	1
1	65
2	97
3	97
4	65
5	1

We intuitively assume that the height changes smoothly as the time changes. This is reasonable; the equation gives a smoother representation of the continuous nature of the flight than the table.

DESCARTES BELIEVED THAT the world was mechanical and that the secrets of the universe could be fully explained by mathematical interpretations. His coordinate geometry was an enormous help in that interpretation. Space and time could then be linked, not only through indefinite, unreliable geometric pictures caught by the spirit of intuition, but through algebra, which had been invented by the Arabs in the ninth century, and which by Descartes's time had been winning the trust of mathematicians for 700 years. The concept of a function would have been natural for examining the space-time relationship, but that would have to wait for Leibniz to introduce it in 1692, when he wrote about tangents to curves. By convention, we now use the notation $y = f(x)$ to indicate that f is a rule that assigns to every value of the number x a unique number y. But the earlier notion of a function was that it simply be an expression built from the operations of algebra and analysis—for example,

$ax + b\sqrt{a^2 - x^2}$ would qualify because it is built from the alge-

braic operations of addition, multiplication, exponentiation, and the extraction of roots. The function concept went through many revisions before 1837, when it settled for Johann Peter Gustav Lejeune Dirichlet's brilliant definition: *y is a function of x, if for every value of x there corresponds a unique value of y.* Dirichlet's definition gave no restriction on how the correspondence is carried out. Descartes did not have such a free definition, but he did associate equations with curves and therefore could investigate how one variable moved with another as easily as points in space moved with time.

The Arrow's Trajectory

In 1647, the Jesuit mathematician Gregory of Saint-Vincent completed his 1,200-page opus, a small part of which examined the Zeno "Achilles" paradox as an infinite sum of a geometric series and computed the exact time and place at which Achilles overtakes the tortoise. He was the first person to do so, introducing the idea of connecting the continuum with infinite division of the number line.

In this atmosphere, calculus was invented and motion was its first application. The problem with motion is that it can vary in speed and direction. When motion does not vary in speed—that is, when it covers equal distances in equal times—its speed is simply the distance covered in a unit of time. But what happens when an object's speed varies with time?

In the new math of coordinate geometry, curves and slopes represented how quickly a certain motion changed. Now, not only could velocity be described, but so too could acceleration. Thus in the late seventeenth century, problems of motion took on lofty new challenges, creating mathematical questions that only calculus would be able to answer: What are the velocity and acceleration of a moving body at any instant of time? Given the acceleration, what is the velocity and distance traveled over any time interval? That answers might be forthcoming to these new formulations of old questions, generated by a totally different way of looking at modeling the natural world, was intensely exciting to the growing number of people interested in science.

It is no accident that two mathematicians—Leibniz, "of middle size and slim figure, with brown hair, and small but dark and penetrating eyes," and Newton, "rather languid in his look and manner, which did not raise any great expectation in those who did not know him"—in Leipzig and England, respectively, discovered the calculus simultaneously. The time was ripe.

MATHEMATICIANS OF THE seventeenth and eighteenth century—Newton, Leibniz, Wallis, the Bernoulli brothers, Euler, d'Alembert, and others—were freed from the classical Greek insistence on mathematical rigor, the hallmark of mathematical practice since Euclid perfected the axiomatic system of proof almost 2,000 years before. They were empowered by their intuition and speculation about the infinitely large and the infinitely small. These attitudes sowed and watered the seeds of the infinitesimal calculus, permitting it to grow freely and be nourished without a logically sound root system.

These mathematicians invented new rules and new notations to manipulate infinity as a magician would a never-ending deck of cards. Their definitions were nebulous; their methods hazy, and their logical arguments were compromised by broken links. In the words of Tobias Dantzig, "Intuition had too long been held imprisoned by the severe rigor of the Greeks. Now it broke loose, and there were no Euclids to keep its romantic flight in check." By the sixteenth century mathematics had begun using certain tools of the infinite and infinitesimal along with an intuitive grasp of the continuum. The irrational was being accepted as a number, along with zero and negative numbers even imaginary numbers were beginning to make their way into the vocabulary. Letters were being used as symbols and algebra was being revived from its ninth-century beginnings as a promising branch of learning. The use of algebra and its astute use of symbols prepared mathematics for the calculus revolution.

The essential ideas of calculus had been incubating for centuries, ever since Archimedes was able to make astoundingly excellent approximations of π. Archimedes was also able to obtain the area of a parabola segment subscribing a triangle of area

A as 4/3 A by using the sum $1 + \frac{1}{4} + \frac{1}{16} + \frac{1}{64} + \ldots + \frac{1}{4^{n-1}}$ and

noticing that it approached 4/3. But he did not let n approach infinity, and never defined 4/3 as the sum of the infinite

series $1 + \frac{1}{4} + \frac{1}{16} + \frac{1}{64} + \ldots + \frac{1}{4^{n-1}} + \ldots$, and hence stopped

short of defining the limit concept. Nevertheless, he certainly had the sense that the sum had a potential to be carried out for as many terms as desired and understood that the difference between the sum and 4/3 would shrink as the number of terms increased. In this he came closer to the calculus than anyone before the seventeenth century. Had he understood the problem in the context of limits, involving the spirit of infinite sums, he would have addressed one of Zeno's deepest concerns—that, yes, an infinite sum of positive terms can be finite.

This concept of limit has transformed our understanding of motion. Take the example of the simple relationship $v = t^2$ and suppose that v represents the velocity of an object in units of feet/second moving along a flat surface at time t in the first 4 seconds of travel. The graph representing the motion of the object is the set of points in the plane (p. 121) with addresses (t, t^2). Note that these points (t, t^2) have nothing to do with the position of the object; they merely represent an orderly correspondence between the number of seconds that have passed since the object started to move and the velocity of the object at that number of seconds.

The area under the graph would be reasonably close to the sum of the areas of the rectangles. The area of any rectangle is its base times its height. So the calculated units of our area is

$\left| seconds \left(\dfrac{feet}{second} \right) \right.$, which simplifies to *feet*. Though this is not a

justification for concluding that the area under the graph over the *interval* from 0 to 2 is the distance it travels in the first 2 seconds, it does encourage the idea that it is true. We shall see that the object indeed will travel 8/3 feet in the first 2 seconds.

We may approximate the area under the curve by drawing n rectangles under the curve. The total area of those n rectangles is

$\left(\dfrac{2}{n} \right)^3 \left(1^2 + 2^2 + 3^2 \ldots + (n-1)^2 \right) \cdot$ Using a bit of algebra and

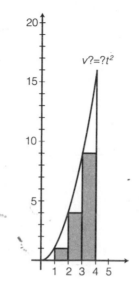

the fact that the sum of the squares of the first n integers is

$\dfrac{n(n+1)(2n+1)}{6}$, we find that the area of the n rectangles reduces

to $\dfrac{4}{3}\left(1-\dfrac{1}{n}\right)\left(2-\dfrac{1}{n}\right)$. From here it is easy to see that the area approaches 8/3 as n grows very large. Is the area under the graph actually 8/3? The modern calculus view is that we may define the area to be whatever the approximating area converges to as long as it is converging to some unique real number. In this case, the approximating area is converging to 8/3, so the area under the curve is actually 8/3 by definition of the area under the curve from 0 to 2.

The point of defining the limit as the value to which the sum is approaching is critical to calculus. We started with a very large collection of numbers that were to be added. We had a method for building larger collections based on a general rule—to get the next larger collection, double the number of rectangles under the graph. What we did—and what medieval mathematicians did not do—was to realize that the infinite sum had a precise meaning as a unique real number. Furthermore, we implicitly envisioned the original graph as the infinite collection of tops of the inscribed rectangles. That shows that after 2 seconds, the object has traveled 8/3 feet. We found the area under the curve over the time interval from 0 to 2. But we could have just as easily found the area over the interval from 0 to t for any value of t between 0 and 4. If we had, it would have turned out to be $t^3/3$ feet.

So suppose we started with the relationship $s = t^3/3$, where s represents the distance traveled in t seconds, where t can be any time between 0 and 4 seconds. Suppose further that we want to know how fast the object is moving when the clock strikes 1 second. We look at the curve (p. 123) and find that its coordinates are in units of time versus distance, and hence the horizontal unit is seconds and the vertical unit is feet. We notice that when the curve is steep, the speed is great.

This suggests that the speed may have something to do with the steepness of the curve. But the slope of the curve measures steepness; that is, by the ratio of the vertical and horizontal components, which are in units of distance (say feet) and time (say seconds). The units of steepness is therefore $\dfrac{feet}{second}$, which is a unit of speed. Once again, this is not a rigorous proof, but it does encourage the idea that the slope of the graph over

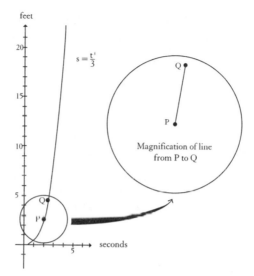

$t = 2$ will give us the speed of the object at the instant $t = 2$.

Now let us see how we might go about finding that steepness. Imagine that the curve is not quite so curved. If the neighborhood of the graph around the point P is enlarged enough, it begins to look very straight—the bigger the enlargement, the straighter the graph. Near the point P with coordinates $(2, 8/3)$, there is another point Q with coordinates $\left(2+h, \frac{(2+h)^3}{3}\right)$. In other words it is a point over $t = 2 + h$, where h is some small number. If h is small enough, Q will be in our enlarged neighborhood. The steepness of the line through P and Q may be represented by the ratio of the height and base of a right triangle with hypotenuse sitting on the line segment between P and Q. That ratio

is $\dfrac{\left(\dfrac{(2+h)^3}{3h} - \dfrac{8}{3h}\right)}{h}$. With a bit of algebra it simplifies to $4+2h+\dfrac{h^2}{3}$.

We can see that as h shrinks, this expression approaches 4. With a bit more work and more bulky algebra, we could have found the slope at an arbitrary time t. Had we done that, we would have found the slope to be t^2. Notice that this is the function that we first started with. If the distance traveled is described by the relationship $s = t^3/3$, then its speed at time t is just t^2, and conversely, if the speed at time t is given as $v = t^2$, then the distance traveled at time t is $s = t^3/3$.

Notice the symbiosis of the relationship between the speed and distance graphs; either one mathematically determines the other.

THE IDEA OF representing the area bound by a curve as a sum of geometric figures of known area goes back to Democritus in the fourth century BCE. One might even go as far as to say that the idea of calculus starts with the question of how to put a numerical measure on a geometric situation. If this is so, then even Pythagoras should be thanked for his contributions to the subject. But the kinds of questions that interested those in the century before Newton would have had no meaning for Greek mathematicians. A fourth-century Greek would never ask for the area of a circle. He would have a notion of area as the numerical measure of the space inside, but he would have no way of handling such a measure. He would be asking how many square units fit inside the circle. But he would also have known that the square and the circle do not have a common measure, and so there would be no straightforward answer. Instead, he would say that the ratio of the areas of two circles is the same as the ratio of areas of squares on their diameters, avoiding the nasty problem of coming up with a common measure for squares and circles. He would be simply comparing squares with squares and circles with circles. The Greek scheme for finding areas was to reduce all areas to rectilinear figures, those simple geometric figures whose areas were well known. But the circle could not be reduced to a rectangle, so Eudoxus, in the fourth century BCE, invented a scheme for sneaking up on the areas bound by his figures, appropriately named in the seventeenth century *the method of exhaustion*. This method was not new. In the fifth century BCE, Antiphon the Sophist had used a similar process to find the area of a circle in a gallant attempt to square the circle. He simply inscribed a regular polygon (a polygon whose sides are all equal) inside the circle and successively doubled the number of sides, watching the polygon geometrically grow close to the circle. Though we don't know what he had in mind to do with this observation, we may easily surmise that he would stop the doubling at a moment when he felt comfortable enough with computing the numerical value of the area of the approximation. He did not take the critical step to notice that the numerical evaluations of the successive areas were converging to some number, the step that modern calculus would take.

Mathematicians have used the geometric trick to establish areas and volumes of regular figures—Kepler applied it to find that the volume of a sphere is one-third the radius times the surface area of the sphere. He imagined the sphere as a very large number of cones with their vertices converging at the center of the sphere and their bases resting on the surface of the sphere.

The subtle and creative idea behind calculus is that we may

define any smooth curve as the result of an infinite succession of finite polygons, and that our given curve is *not just an approximation by polygons that is good enough* when the number of sides of those polygons is high enough. We have the Flemish mathematician Gregory of Saint-Vincent and his 1,200-page tome to thank for this idea. He was the first to write that an infinite sum can be finite.

But is the limiting value of our infinite sum ever attained? This question provoked lively debate in the eighteenth century; it was at the root of Zeno's Achilles argument. The short answer is no, the sum does not reach its limit. The motion paradox lives on.

When the Italian Bonaventura Cavalieri wrestled with the Achilles paradox, he also found a way to handle the infinite. In 1629 he devised a clever scheme for sidestepping issues raised by Zeno, letting intuition guide his mathematics to generalize the results of Archimedes. We can imagine him asking out loud, "Imagine the solid as though it were simply made from very thin pages of a book; then wouldn't its volume be very close to the sum of the volumes of its pages?" He could see that the area of a moving stick in the plane should simply be the length of the stick times the length of the path traced out by the stick's midpoint, if the path is perpendicular to stick. He would have tested this on a stick of length r rotating about one of its endpoints. It would sweep out a circle of radius r while its midpoint would trace a circle of radius $r/2$. Hence, according to his principle, the area swept out would be the circumference of the circle of radius $r/2$ times the length of the stick. This translates into mathematics as $2\pi\left(\dfrac{r}{2}\right)$ times r, or πr^2, the true area of a circle of radius r. How wonderful! No proof, yet it seems to make so much intuitive sense.

Nicole Oresme, back in the fourteenth century, had already constructed the relationship between velocity and distance by making connections to areas, crudely hinting at what would later be known to calculus as the *integral*. Galileo suggested the ideas of the infinitesimal and the continuum—ideas that now lie at the heart of calculus—at the very beginning of his *Dialogues Concerning Two New Sciences*, but quickly dismissed such notions as incomprehensible in the light of multiple paradoxes springing from infinity, though he proceeds to use those ideas in the Third and Fourth Days of his *Dialogue*.

Another Italian of the seventeenth century, Torricelli, cleverly exhibited an infinitely tall figure with finite area. He also showed that the volume of the infinite surface generated by revolving a portion of a hyperbola about one of its asymptotes is finite. This may have seemed paradoxical at the time, since it would seem that one could paint the infinite surface by just pouring a finite amount of paint inside and then pouring the paint out. Modern-day calculus students know this trick. It comes to us by way of attempting to answer Zeno's dichotomy: Yes, an infinite sum can be finite.

Galileo's clever idea was to take the ratio distance/time covered in a small interval of time and to use the obvious fact that the resulting ratio depends on the size of the interval. He built a sequence of such ratios with the time interval becoming smaller and smaller, noticing that as more terms of the sequence were built, the last term would come closer to some specific value. For example, if the distance the object travels in t seconds is t^2, keep dividing the time interval between 1 and 2 in half and consider the ratios of distance to time. They are $\dfrac{3}{1}, \dfrac{5}{2}, \dfrac{9}{4}, \dfrac{17}{8} \cdots$

These are the average speeds in the respective time intervals. It is as though the speed were constant in these intervals. Now Newton would say that if the lengths of the broken time intervals continue to shrink indefinitely, then the sequence that results will represent the object's speed. In other words, for any given instant of time t there will be an interval including t. The speed at the instant t is defined to be the average speed in that interval. Intuitively, there is nothing wrong with such a definition of speed, but where is the rigor?

To get ourselves on firmer ground we would ask for the criterion by which we would know how small to shrink the time interval and how close we should come to that specific limiting value we call speed. Today we have criteria that are defined by arithmetic definitions and axioms not available to Newton. And yet, Newton was able to formulate extraordinarily fruitful models in his investigation of nonuniform motion.

THE PROBLEMS OF motion were so intimately connected with the mysterious infinite that they could not be separated from questions of the relationship between space and time, and in particular from Zeno's original paradoxes. The mathematics of motion involves grasping infinity, and that notion has always to some extent contradicted human intuition. An infinite sum can be finite? Dimensionless points and breadthless lines also sound absurd—are they not all fictions of the geometer's mind?

Zeno intuitively knew his arguments were consistent, though rules of logic had not yet been invented; indeed, he could have predicted that infinity would one day tightly come under mathematical control, but he was also wise enough to know that it would remain jarring to human sensibilities.

Zeno pulled an arrow from his quiver, shot it from his bow, and asked us to stop time to examine a stationary arrow without destroying its flight. The mathematician can do that easily—stop time and abstractly visualize the arrow—and believe that the frozen arrow is indeed one and the same as the one shot. But he or she is simply replacing a mathematical abstraction with a mental impression of a fixed arrow—one that may even be clearly visualized as if on a screen in the mind—but it is not the real arrow that smoothly moves from its bow to quiver in its target.

What makes it paradoxical is the apparently smooth flow of time. But we must regard that smoothness as an assumption rather than the truth. It is not the job of mathematics to verify whether time is really continuous or not—mathematics doesn't really care. Even if we completely accept calculus's most up-to-date definition of limit, we still would not have explained Zeno's dichotomy paradox. Though the paradox can be stated in precise mathematical language, the dichotomy is really a consequence of the natural dynamics of motion, which is a matter of physical phenomena and the way our minds work, not math.

The mathematical explanation of the dichotomy paradox is simply the statement that the infinite sum

$$\frac{1}{2} + \frac{1}{4} + \frac{1}{8} + \ldots + \frac{1}{2^n} + \ldots$$

equals 1. We demonstrate this as follows: let

$S_n = \dfrac{1}{2} + \dfrac{1}{4} + \dfrac{1}{8} + \ldots + \dfrac{1}{2^n}$ represent the sum of the first n

terms. Then $\dfrac{1}{2} S_n = \dfrac{1}{4} + \dfrac{1}{8} + \dfrac{1}{16} + \ldots + \dfrac{1}{2^{n+1}}.$ Subtract $\dfrac{1}{2} S_n$

from S_n to get $S_n - \dfrac{1}{2} S_n = \dfrac{1}{2} - \dfrac{1}{2^{n+1}}$. Hence

$\dfrac{1}{2} S_n = \dfrac{1}{2} - \dfrac{1}{2^{n+1}}.$ As n grows large, the term on the right

shrinks toward 0. We say that $\dfrac{1}{2} S_n$ approaches $\dfrac{1}{2}$ and hence

that S_n approaches 1. We seem to have shown that

$$\dfrac{1}{2} + \dfrac{1}{4} + \dfrac{1}{8} + \ldots + \dfrac{1}{2^n} + \ldots = 1.$$

It is strange. We accept the details of the algebra, and sud-

denly we have jumped from the shrinking of $\dfrac{1}{2^{n+1}}$ to accepting

that the infinite sum equals 1. We are already implicitly using the notion of *limit*.

The Achilles paradox is largely a matter how we interpret the race. Calculus views it as a question of Achilles "catching up" with the tortoise, which is confusingly different from Achilles overtaking the tortoise. The idea of "catching up" is modeled by the mathematical notion of limits, which requires Achilles to persistently "get closer to" the tortoise without ever reaching it.

Suppose that Achilles's speed is 10 miles per hour, the tortoise's speed is 1 mile per hour, and the tortoise is given a head start of 9 miles. Then, when Achilles runs 9 miles, the tortoise is 9.9 miles from the starting line. When Achilles is 9.9 miles from the starting line, the tortoise is 9.99 miles from the starting line. At Achilles's n-th attempt to catch up, the tortoise will be 9.99 . . . 9 miles from the starting point, a number having n nines after the decimal point. If the race continues indefinitely, the tortoise will be 9.99 . . . miles from the starting point. The dots at the end of this last expression indicate that there is an infinite string of nines after the decimal point. This last expression has infinitely many digits; so we have to make some sense of what adding infinitely many digits could possibly mean.

Suppose the number 9.99 . . . 9 had n nines after the decimal point, where n is an arbitrary large number, any large number. Now, let n increase and watch what happens to the number as n increases. If it approaches some number, say 10, then it seems reasonable to say that the infinite expression is really representing the number 10. The more nines there are, the closer 9.99 . . . is to 10; so we say that 9.99 . . . = 10. We can do this simply by defining 10 as the infinite string of 9s, because the infinite string of 9s has never been defined. In other words, we must agree that the definition of the infinite sum $9 + \dfrac{9}{10^1} + \dfrac{9}{10^2} + \dfrac{9}{10^3} + \cdots$ is 10.

We have another way to look at the problem. Notice that Achilles travels $10t$ miles from the starting line in t hours and that $9 + t$ is the distance that the tortoise travels from the starting line in t hours. That is because Achilles travels 10 miles per hour for t hours; and the tortoise will have traveled $9 + t$ miles, because he had a head start of 9 miles and was traveling at only 1 mile per hour.

At the moment Achilles catches up with the tortoise, we know that the distance from the starting line for Achilles must be the same as the distance from the starting line for the tortoise, and so, $10t = 9 + t$. Hence, Achilles will catch up with the tortoise when $t = 1$ hour. And the distance traveled from the start by Achilles will therefore be 10 miles.

This is the usual answer that mathematicians give when confronted with the Achilles and tortoise paradox. But how will Achilles be able to perform an infinite number of things in a finite amount of time? This difficulty disappears as long as we do not confuse motion in space with the movement of time.

The arrow paradox also requires an understanding of limits as a mathematical model for instantaneous velocity, which calculus treats as a *derivative*, an instrument that creates limits of average changes of a dependent variable in small intervals of an independent variable. The model here is to view each point on the arrow's trajectory as though it were a limit of a sequence of rational numbers on the number line, so the arrow's path is assured a persistent even flow of space in the continuity of time. In effect, it assumes, quite correctly, that all numbers on the number line are convergent sequences of rational numbers.

But an understanding of continuity is still elusive. Leibniz had thought that continuity was satisfied by any collection of numbers with the property that between any two elements there is a third in the collection. By his criterion, the set of rational numbers is continuous. But it is not, for $\sqrt{2}$ presents a gap. The stadium paradox informs us that motion must be continuous in a continuous space. Mathematics can make good definitions of what it means to be continuous, but it cannot decide on whether or not physical space is continuous, nor can it decide whether or not motion in that space is continuous. As the mathematical historian Carl Boyer once remarked, "The paradoxes of Zeno are consequences of the failure to appreciate this fact."

Gregory of Saint-Vincent deconstructed the Achilles paradox as a limit of an infinite series, an explanation very close to modern reasoning. He asked what would happen if subdivisions of the peculiar race between Achilles and the tortoise were permitted to continue indefinitely, and found that such a race could be modeled as an infinite geometric series. For Gregory, the paradox was simply a question of summing the infinite series. Galileo argued that there is some proportion between the speeds of the two competitors and calculated, by geometric progression, the point at which Achilles would meet up with the tortoise, ignoring the phenomenological reason for why it should ever happen. No one doubts that some mathematical model could pinpoint the place where Achilles meets the tortoise, but Gregory, using geometric series, was the first to state the exact time and place when it will happen. The problem is *how* does it happen when the Zeno paradox is dictating that it cannot? We must be able to explain the reason it happens without resorting to a model that may or may not precisely represent what we observe. After all, Achilles is not a point and neither is the tortoise.

THE GHOST OF Zeno appeared in 1734 as George Berkeley, the Anglican Bishop of Cloyne in County Cork, Ireland. In Berkeley's view, calculus, the new mathematics, failed to conform to intuitive notions of continuity and had been built on weak foundations. The subtitle of his essay *The Analyst* gives his point of view: *Or a Discourse Addressed to an Infidel Mathemati-cian. Wherein It Is Examined Whether the Object, Principles, and Inferences of the Modern Analysis* [meaning calculus] *Are More Distinctly Conceived, or More Evidently Deduced, than Religious Mysteries and Points of Faith. "First Cast the Beam Out of Thine Own Eye; and Then Shalt Thou See Clearly to Cast Out the Mote Out of Thy Brother's Eye."* The real argument was over the jus-tification of Newton's ambiguous meaning of limits of ratios where both numerator and denominator tend toward zero, a misleading notion that ignored all the appreciation of the subtle nuances and difficulties of infinity and continuity. To Berkeley, it seemed zero divided by zero was a meaningless contradic-tion.

The Bishop was right, just as Zeno was. Intuition is fine for those with good intuition: Euler, Fermat, Newton, and Leib-niz. The danger was that something slyly anarchic could slip through the front gates of calculus disguised as the legitimate heir to a proven theorem. By the end of the eighteenth century, mathematical contradictions were multiplying. Yet practical applications of calculus and coordinate geometry were explod-ing, improving human lives and knowledge of the real world without regard to the inconsistencies sneaking through the gates of reason. Despite those inconsistencies, this math blos-somed for more than two centuries, ignoring its loose footing on logic, with remarkably few serious errors.

Mathematics developed hand in hand with practical needs, developing new fields as they emerged. Questions about motion turned into abstract questions about velocity, tangents to curves, maximum values, and lengths of curves. The answers, even preliminary tentative ones, gave tremendous insight into practical problems. Fermat, Newton, and Huygens studied the angle at which a ray of light bends when passing through a lens, which requires knowing the tangent to the surface of the lens. Others applied this new math to warfare, such as the range of a cannon, which depends on the initial angle of flight of the cannonball. Euler and Lagrange set up the wave equations for the propagation of sound; Daniel Bernoulli analyzed the tones given off by musical wind instruments, while Jean-Philippe Rameau was writing for harpsichord and flute. Interest in new musical instruments prompted work on understanding the vibrating string after Bach adapted the concerto principle to keyboard and orchestra. In 1747 and 1748, Euler and d'Alembert made many contributions to the mathematical theories of music, including an understanding of the waves of a vibrating drum, while studying the motion of strings on musical instruments.

IT CAN BE argued that coordinate geometry and the invention of calculus opened up the study of architecture, astronomy, artillery, carpentry, cartography, celestial mechanics, chemistry, civil engineering, clock design, hydrostatics, hydrodynamics, music, optics, pneumatics, ship construction, thermodynamics, magnetism, materials science, and navigation—and this list is not exhaustive. Together they constitute one of the greatest revolutions in the history of mathematics. But still, the paradox of motion was not fully answered.

When Newton gave figurative credit to those giants on whose shoulders he stood, he must have been referring to René Descartes, for paving the way with coordinate geometry; and

Evangelista Torricelli, Bonaventura Cavalieri, Gilles Personne de Roberval, Pierre de Fermat, Blaise Pascal, and John Wallis, who independently used the so-called method of exhaustion to extend the work of Archimedes on areas of spirals; as well as Gregory of Saint-Vincent, Newton's teacher Isaac Barrow, and others lost to history. But that list does not include men like Kepler, Galileo, Copernicus, and others, those who helped guide him to his most famous accomplishment: the marrying of his beautiful inverse square law to his notion of gravity. What is to be learned from it about the nature of motion has only begun.

· 10 ·

Falling Toward the Enlightenment

Sixty-five years after Shakespeare wrote *The Tragedy of Hamlet*, Newton conceived of the law of universal gravitation, which—in addition to having many enormous consequences for physics—relaxed the idea that man's fate was linked to the movements of the heavens. Now falling apples and the attraction between planets were surely linked; man's fate and the movements of the stars were not.

Translated around the time Hamlet first walked on stage, the King James Bible claimed that "The sun also riseth, and the sun goeth down, and hasteth to his place where he arose. The wind goeth toward the south, and turneth about unto the north; it whirleth about continually and the wind turneth again according to his circuits. All the rivers run into the sea; yet the sea is not full; unto the place from whence the rivers come, thither they return again." For Aristotle, everything in the universe had a natural place to which it would strive to return when moved. But by the late eighteenth century, gravity was beginning to be thought of as a possession of systems of matter: Two objects attract because they are a certain distance apart and they contain a certain amount of matter, by virtue of their "bulk" and motion. What scientists of the late eighteenth century called *force* we call either momentum or kinetic energy. Newton thought of gravitational forces as dependent on their relations with other bodies. A body in isolation has no intrinsic gravitational force, but when another body comes near, it exerts force on that body, and that body exerts a force back.

The sixteenth century was still laboring, as Aristotle had, to articulate universal laws, but by the eighteenth century the prevailing scientific view was that law determines the universe. Yet unlike the motion of the planets, the governing laws of biology are dependent on far too many variables to be perfectly explained. An apple may fall from a tree and abide by Newton's simple laws of motion, but the apple itself is an extremely complex bundle of molecules held together by a formidable number of complicated internal atomic pulls.

In Milton's *Paradise Lost*, God sends the archangel Raphael down to Paradise to admonish Adam and also to uncover the identity of Satan. Raphael is entertained at a table "with pleasant liquors," the finest fruits and meats of Paradise brought by Eve, while Adam asks about the world, how it came to be, and how the planets move. Raphael explains:

> . . . Heaven
> Is as the Book of God before thee set,
> Wherein to read his wondrous works, and learn
> His season, hours, or days, or months, or years,
> This to attain, whether Heaven move or Earth, . . .

Hereafter, when they come to model Heaven
And calculate the stars, how they will wield
The mighty frame, how gird the sphere
With centric and eccentric scribbled o'er,
Cycle and epicycle, orb in orb. . . .

Milton completed *Paradise Lost* just before the Great Plague hit London in 1665, when Newton left Cambridge and took refuge at his childhood home in the hamlet of Woolsthorpe, where he discovered, among other things, his universal law of gravitation, the description of the composition of the action of gravitational force with inertial motion that both holds the planets in their orbits and causes the apple to fall.

To see the planets move along their elliptical orbits, a person would have to be fixed relative to the sun, presumably watching from a position high above the center of the sun. How does a person arrive at conceiving that the planets travel uniformly in elliptical orbits, when the picture we see in the sky is of planets dancing like cowboys lassoing the stars? (See time-lapse photo of planets taken over a twenty-year period by the Munich Planetarium on p. 140.)

It makes no difference whether or not we take the earth as fixed or the sun, or any other part of the universe. Ptolemy was just as *right* as Copernicus. Fortunately, each model has its consequences. Copernicus's fixed-sun model is not only simpler than one that involves dancing planets, but one that can be explained by a universal law. And if the paths of planets are to be ellipses, the law simply describes the composition of motion under the action of gravitational force with inertial motion (the motion a body has in virtue of its mass). The beautiful thing

Deutsches Museum Archives, Munich

about the mass of a body is that it is independent of where an object is; it, by itself, does not care how close the body is to the gravitational field of another body. This means that the acceleration (change in speed) of a body only depends on the forces exerted on the body. So the body's motion is entirely a mathematical problem, with Newton's wonderful law of universal gravitation (the inverse-square law) as its mathematical agent.

When we talk about who moves around whom, we are talking about a path of motion, not about a law of movement. One path of motion is equivalent to another when looked at from the right relative moving reference point. But once we choose which planet or star is fixed, we no longer have the choice of law. If the inverse-square law works for apples and cannonballs, then we should love it to work as well for planets and the moon. Though a sun-centered solar system is consistent with both observation and the inverse-square law, an earth-centered system would require a nasty coordinate change and far more complicated geometry to accommodate Newton's wonderful law. Copernicus was right, after all—if by *right* we mean something like nifty and elegant.

One may wonder how Newton thought of his wonderful law and of why he thought the force to be—out of all the many possibilities—inversely proportional to a square. But the square was not an innovation. Pythagoras used squares to express his magnificent relationship between the sides and hypotenuse of a right triangle; Apollonius used them to describe defining characteristics of conic sections; the Merton College kinematicists used them in formulating their acceleration theorem, the first truly mathematical formulation of a law of motion; Kepler used them in his third law, connecting the time a planet takes to make one complete revolution to the average distance from the sun.

Leibniz used them in a 1686 paper, insisting that the energy of a moving body is proportional to the *square* of its velocity—not linearly proportional to velocity, as Descartes had had it. Leibniz's idea was imprudently rejected until the brilliant, beautiful, and rich Gabrielle-Emilie de Breteuil, Marquise du Châtelet revived it in 1746. She interpreted the works of Descartes, Leibniz, and Newton, and performed experiments in her splendid laboratory to observe the physical consequences of their mathematics and empirically demonstrate that energy is proportional to the square of velocity by dropping brass balls from 1, 4, and 9 units of height above a clay bed to leave craters of depths 1, 2, and 3, respectively.

• • •

IN THE SEVENTEENTH century, there was a considerable interest in the motion of the moon, the motion of the planets, the
motion of objects falling to Earth, and the motion of projectiles
fired from cannons, which were first used by the English in the
Battle of Calais in 1347 during the Hundred Years' War. However, beside ethereal winds or the theory of Descartes's vortices—based on the notion that the universe is filled with an
invisible liquid in which the planets are carried around by
whirling vortices—it was still not fully known what caused the
elliptical orbits of the planets with the sun at one focus. For
Descartes, all space is full of matter, so particles must move in
circuits and his mechanics is really kinematics; there is no acceleration or deceleration. So the only way "force" is involved is
that momentum (the motion of a body by virtue of its mass) is
conserved before and after impact.

Herbert Turnbull, the twentieth-century mathematician and
Newton historian, tells this delightfully fanciful story about the
young Newton:

> In the country near Grantham during a great storm, which
> occurred about the time of Oliver Cromwell's death, a boy
> might have been seen amusing himself in a curious fashion.
> Turning his back to the wind he took a jump, which of
> course was a long jump. Then he turned his face to the
> wind and again took a jump, which was not nearly so long
> as his first. These distances he carefully measured, for this
> was his way of ascertaining the force of the wind. The boy
> was Isaac Newton, and he was one day to measure the
> force, if force it be, that carries a planet in its orbit.

The story of how gravitation was discovered may be as apocryphal as the one about a falling apple, but it goes roughly as follows.

Christopher Wren was the architect of St. Paul's Cathedral in London; Edmund Halley was a renowned astronomer; and Robert Hooke a physicist. The three men would often meet to discuss topics ranging from the taste of beer to the meaning of life—physics and mathematics thought to be somewhere in between. One day in 1684 they met in London and talked about Descartes's idea that the movement of planets was caused by regular and stable winds blowing the planets around vortices in an invisible liquid. After a long discussion, they hit on the idea that it may not be Cartesian winds, but rather some sort of gravitational force that keeps them in orbit. Perhaps the sun itself kept the planets in orbit, by some miracle forcing deflection of motion from a straight line.

"What if," asked Wren, "the sun pulled the moving earth by some strange pulling force. Would that pull dictate an elliptical orbit?"

"Ah," said Halley, "you are suggesting that if we can mathematically show that the sun's pull forces an elliptical orbit, then we would be reasonably sure that it is the sun that does the pulling."

"I can answer that," said Hooke.

"I'll give you forty shillings if you do by next month," said Wren. (Forty shillings in 1684 was equivalent to about a thousand dollars today.)

But Hooke did not return with an answer.

Several months went by before Halley met with Newton in Cambridge and asked his opinion on the matter.

"Isaac," said Halley, "if the sun were to pull a moving planet with a force inversely proportional to the square of the distance between it and the planet, what type of curve would the planet follow?"

"An ellipse," Newton answered without the slightest hesitation.

"How do you know?" Halley asked, astonished.

"Why, I have calculated it."

"May I see the calculations?"

"I'll have to find them. When I do, I'll send them to you."

A theory—or, really, a suggestion—about the sun causing the planets to move in elliptical orbits seems a good first step. But when was the notion of *universal* gravitation put forward?

Then, of course, we have the legend of the falling apple. The original comes from Newton's close friend William Stukeley, who recalled what happened after spending a day and dining with Sir Isaac at his Kensington lodgings on April 15, 1726:

> After dinner, the weather being warm, we went into the garden and drank tea under the shade of some apple trees, only he and myself. Amidst other discourse, he told me he was just in the same situation as when formerly the notion of gravitation came into his mind. "Why should that apple always descend perpendicularly to the ground?" thought he to himself: occasioned by the fall of an apple, as he sat in a contemplative mood. "Why should it not go sideways, or upwards? But constantly to the earth's centre? Assuredly, the reason is that the earth draws it. There must be a drawing power in matter and the sum of the drawing power in the matter of the earth must be in the earth's center, not in any side of the earth. Therefore does this apple fall perpendicularly, or toward the center? If matter thus

draws matter, it must be in proportion of its quantity. Therefore, the apple draws the earth as well as the earth draws the apple."

The idea of the earth's gravitation was not new, but Newton's notion that "the apple draws the earth" was radical and far-reaching. In this view, what brings the apple down holds the moon in its path as though it were like any other projectile. What holds the distant moon is what pulls the nearby apple— or a faraway planet. Suddenly the entire universe is filled with bodies, large and small, pulling each other in all directions, everything pulling everything, apples pulling planets, planets pulling apples. From this, Newton was able to relate the force of attraction between two objects to their masses and the distance between them.

When Halley visited Newton in Cambridge and asked what type of curve a planet would follow if it were tethered to the sun by a force inversely proportional to the square of the distance between the two, Newton immediately claimed it to be an ellipse. He knew because eighteen years before, while working in the quiet of his little manor house isolated from the raging plague, he had considered the inverse-square law.

Newton stumbled on a miracle. With a great deal of skepticism based on very crude observations, he formulated the law of universal gravitation as a remarkably simple mathematical creation that turned out to give amazingly accurate descriptions of motion. He worked on the calculus between 1665 and 1677, but did not publish it. In 1676, after learning that Leibniz was working on something similar, he sent Leibniz two letters cryptically presenting his thoughts on calculus, avoiding hints of his methods and earnestly acknowledging his indebtedness to others. The English mathematician John Wallis, who himself contributed substantially to the origins of calculus, wrote, "Mr. *Isaac Newton*, the worthy Professor of Mathematics in *Cambridge*... about the Year 1664, or 1665 ... did with great sagacity apply himself to that Speculation [of Infinite Series]. This I find by Two letters of his (which I have seen,) written to Mr. *Oldenburg*, of that Subject, (dated *June* 13, and *Octob.* 24. 1676,) full of very ingenious discoveries, and well deserving to be made more publick. ..."

Newton published his ingenious discoveries in his great work *Philosophiae Naturalis Principia Mathematica* as a collection of geometric propositions on velocity and acceleration expressed in geometric terms, which would look uncomfortably imprecise by today's standards.

The events leading to the publication in 1687 of his great work—referred to more briefly as the *Principia*, the book that introduced the laws of motion as a foundation to encompass mechanics of the planets, hydrodynamics and waves, orbits of comets, and the theory of tides, and that transformed mechanics into an exact science—were recounted by two of his close friends at the Royal Society, John Conduitt and William Stukeley. One prevailing story has it that there was a discrepancy between the force necessary to keep the moon in orbit and the inverse-square law applied to earth and moon. But Newton himself reminiscing late in life did not talk of any serious discrepancy:

I deduced [from Kepler's third law] that the forces which keep the planets in their orbs must be reciprocally as the squares of their distances from the centers about which they revolve: and thereby compared the force requisite to keep the moon in her orb with the force of gravity at the surface of the earth, and found them answer pretty nearly. All this was in the two plague years of 1665 and 1666, for in those days I was in the prime of my age for invention, and minded mathematics and philosophy more than at any time since.

He found "them answer pretty nearly." That doesn't sound like a discrepancy. Newton tossed the idea aside because of another difficulty. He had to show that the gravitational pull between two bodies is the same as a pull between their centers of gravity would be. This would have been a dreadfully difficult thing for the young Newton to prove.

The problem for Newton was that, when it comes to masses the size of the earth or moon, the distance between them becomes ambiguous. When applying the inverse-square law, should he take the distance between masses to mean the distances between their surfaces or centers? Perhaps the most intuitive answer is their centers, but this has a mathematical rationale, too.

The law of universal gravitation is intimately connected to calculus through a subtlety: The law says that if F equals the force of attraction between two masses r units apart, then the product r^2F does not depend on either F or r.

The law is commonly expressed as saying that the force of attraction between *any* two bodies is inversely proportional to the square of the distance between them. The mathematical formulation says that if the masses are known to be m_1 and m_2, then the force between them is given by $F = G\dfrac{m_1m_2}{r^2}$, where r

is the distance between their centers of gravity. The magic of this formula is tinctured in the letter G, which stands for a universal constant, a number that does not depend on any particular place in the universe. It is the same for the masses of cannonballs and Earth as it is for Jupiter and Earth.

He argued that the moon would continue with a uniform speed in a straight line if it were not continually pulled by gravity to curve toward earth. It was well known that cannonballs fired from cannons traced parabolic curves. He reasoned that if a cannon could fire a cannonball a great distance, say from London across the English Channel to France, the trajectory would still follow the path of a parabola. So he imagined what would happen if the cannonball were fired with such tremendous force as to reach India. Or China. Or Japan. What would happen if the firing were so powerful that the cannonball went beyond Japan? He reasoned that it should miss the earth entirely but continue falling toward the earth, surrendering to an eternal circular orbit about the planet.

The inverse-square law was just a part of the idea behind universal gravitation. Ever since Galileo performed his experiments with motion, it was known that an object in motion would continue to move in a straight line unless it was compelled to change that motion by some external force. Newton took this as a starting hypothesis for his *Principia*. With this assumption, he could explain the elliptical orbits of the planets by arguing that planets are drawn away from straight line motion by tugs-of-war with distant masses.

Newton knew from Kepler's laws that planets move in elliptical orbits, but he also knew that those elliptical paths were not too far from circular, and if he could work out the details of the simpler circular scenario, he might have a good handle on the general case. So he first considered how planets accelerate if they followed uniform circular motion along a circle of radius r and found that the acceleration (instantaneous change in velocity) a is related to velocity v through the equation $a = \dfrac{v^2}{r}$, which gives acceleration in terms of both v and r. This seemed to mean that the acceleration depends on r and v.

But if T represents the period of the orbit of, say, Mars about the sun—that is, the time it takes for Mars to complete one Martian year—then $T \cdot v$ is the distance the planet travels in one complete cycle. And if the orbit is truly circular, then $T \cdot v = 2\pi r$. Dividing both sides of this last expression gives $v = \dfrac{2\pi r}{T}$.

Squaring both sides of this last equation gives $v^2 = \dfrac{4\pi^2 r^2}{T^2}$. Substituting this representation of v^2 into the equation $a = \dfrac{v^2}{r}$ gives $a = \dfrac{4\pi^2 r}{T^2}$. The acceleration still does not purely depend on an inverse square of the distance, for it now seems to depend on T as well as r. Alas, it seems that after all this work, dependence on v was traded for dependence on T. But one of Kepler's planetary laws says that the square of T is proportional to the cube of r; i.e., $T^2 = C \cdot r^3$, where C is some number that does not vary. Substituting this representation of T^2 into the formula for acceleration $a = \dfrac{4\pi^2 r}{T^2}$ gives $a = \dfrac{4\pi^2}{C} \cdot \dfrac{1}{r^2}$. Acceleration is then proportional to the inverse square of the radius of the hypothetical circular orbit of the planet about the sun. The inverse-square law then follows Newton's second law of motion, because acceleration is proportional to force.

So, r is the distance between centers of gravity, and that is easy to see if we are talking about the sun and the earth. But what happens at, say, Mount Fuji? Don't the mutual forces between inner parts of the earth—those extremely complex bundles of formidably complicated internal forces that hold all the constituent parts of inhomogeneous earth together—contribute forces to vastly complicate the final resultant forces: the dense rock of Mount Fuji with, say, the island of Madagascar, or the water of the deepest, darkest depths of the Pacific? Newton brilliantly invented and used calculus to solve this problem by showing that the center of gravity takes those internal forces into account, simplifying the form of the law of universal gravitation by concentrating the mass at a single point. Calculus first sees each bulky mass (the earth, for example) as an ordered collection of small masses, each mutually affected by the inverse-square law. It then uses its powerful limit arguments to converge on the collective effect. Even though a mass may be bizarrely irregular and vast in size, the force it contributes acts as though it is coming from a single, mathematical point at its center of gravity. This simplicity is its beauty.

THE TELESCOPE HAD been invented and perfected and the seas around the world had been explored long before Newton was born, and still, witches were being hanged or burned; traitors and criminals were routinely being beheaded in public squares—their heads parboiled for preservation and hung from posts along busy streets—and alchemy continued in the face of the new science of chemistry.

By the time of Newton's death in 1727, eyeglasses and newspapers were readily available and affordable. Enormous political changes had enveloped Europe; small ducal states of central Europe had begun condensing through wars and mergers to become kingdoms, while neighbors shaved large regions from Poland and the Ottoman Empire. Populations of cities remained small—London had fewer than 600,000 inhabitants, Paris fewer than 700,000—and wolves still roamed freely outside the cities. Brightly lit coffeehouses with comfortable furniture and luxurious surroundings were everywhere in the big cities of Europe as well as in university towns, where newspapers such as the *Daily Courant* or *London Gazette* were sold each afternoon and streets were lit at night so people could walk about discussing politics, philosophy, and the latest scientific discoveries. Europe was seeing a fresh style of life. Coffeehouses were not just places of gossip and news, but places where students and faculty could talk about the books they read, discuss poetry and plays, collect mail, or hear the latest scientific reports. Scientific academies and societies were established with funds for publishing periodicals and money for developing research tools and costly measuring instruments.

In the fifty years following Newton's death, Denis Diderot would complete seventeen large volumes of the first encyclopedia, Edward Gibbon would shock the world with his *Decline and Fall of the Roman Empire*, Jean-Jacques Rousseau would write *The Social Contract*, James Watt would build the steam engine, Mozart would write serenades and symphonies, Bach would die, and Beethoven would be born.

Though slave trading increased and wars involving countries all over Europe continued over colonies, trade, and sea power, science, art, literature, and practical inventions were about to explode in the Age of Enlightenment. A middle class was becoming informed and beginning to think, not only about politics, but also about science and literature.

Global information highways were in place and growing for news of calamity, intellectual fashions, and scientific discovery. The motions of human culture were growing dramatically more sophisticated and would soon lead to greater discoveries, but the motions of the planet, not to mention cannonballs and arrows, seemed to have been essentially determined by calculus. Zeno's paradoxes probably seemed more irrelevant than ever.

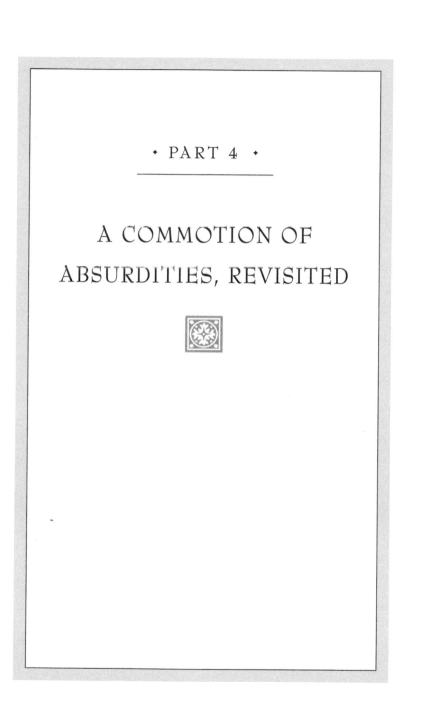

· PART 4 ·

A COMMOTION OF
ABSURDITIES, REVISITED

11

The Speed of Light

 Light would be the key to the next great advance in our understanding of motion. For this, too, we can thank Newton. The eighteen months Newton spent in isolation at the end of London's Great Plague were the richest months of his creative imagination. Working in a quiet, dark room at Woolsthorpe, a room with one shuttered window, a room that doubled as a study and laboratory, the twenty-four-year-old Cambridge graduate experimented with sunbeams piercing through a small hole in the shutter. It was a January afternoon when the sun was relatively low in the sky. A white daylight beam made its way through a triangular glass prism that Newton tactically placed in its path. William Wordsworth later wrote:

> The antechapel where the statue stood
> Of Newton with his prism and silent face,
> The marble index of a mind forever
> Voyaging through strange seas of Thought, alone.

He watched the natural "white" sunlight—which had always been assumed to be absent of color—split into colors of the rainbow. He must have seen light pass through prisms many times before. But the controlled experiment, the isolated environment and his deliberately intense examination of the purely white light on that particular day brought out imaginative ideas on the nature of light. He stood back, hair uncombed, clothes disheveled, staring at the colors, wondering what would happen if a second prism were put in the path of the projected colors. He could not have known the answer. A second prism recombined all the colors back into white.

Others before Newton must have noticed remarkable colors passing through cut glass whenever sunlight passed through, but Newton seems to be the first to have fully explored the question of why the white light entering the prism should exit with a spread of colors always ordered as red, orange, yellow, green, blue, and violet. The first meaningful experiments with light were done by Newton on that January afternoon in 1666; this was the beginning of spectroscopy, the scientific analysis of color and refraction—the study of what ends up to be how we observe motion.

Such an experiment could have been done by almost anyone. Zeno himself. It had long been known that light—which was observed to cast sharp shadows—travels in straight lines, is bent when passing through water or glass, and is turned back by reflection through a mirrored surface. In Euclid's time it was known that light leaves a plane mirror at the same angle it enters. Claudius Ptolemy spent a great deal of time and energy measuring angles of incidence and refraction of light through two different media, wondering why he could not extract a simple law. Such a law would have to wait for the Dutch mathematician and scientist, Willebrord Snell, who in 1621 discovered that the ratio of sines of the angles of incidence and refraction are neatly related to the ratio of densities of the media. Sines, and trigonometry in general, were invented in the fifth century by Hindu astronomers and were thus not available to Ptolemy.

Rainbows had been studied by the early philosophers. Aristotle incorrectly believed them to be caused by reflections of light from droplets of rain. Like most philosophers of his time, he also took the speed of light as infinite. Such a thought should have caused a great clamor of Zeno-like paradoxes of space and time. Sure, the speed of light must have appeared to be astoundingly fast to folks without instruments to measure it—but infinite?

A couple of Arab mathematicians and physicists in the eleventh century, Abd Allah ibn Sina and Abu Ali al-Hasan Ibn al-Haitham, refused to believe that the speed of light was infinite, and even seventeenth-century scientists, including Descartes, believed that light traveled at a finite speed, but it was Galileo who wrote about an experiment to prove the speed of light fast but finite. In his *Dialogues Concerning Two New Sciences*, Sagredo, Simplicio, and Salviati debate. Sagredo asks if the speed of light is instantaneous, "or does it like other motions require time? Can we not decide this by experiment?" Simplicio replies that "everyday experience shows that the propagation of light is instantaneous." He claims that "when we see a piece of artillery fired, at great distance, the flash reaches our eyes without lapse of time; but the sound reaches the ear only after a noticeable interval." To this Sagredo retorts, "the only thing I am able to infer from this familiar bit of experience is that sound, in reaching our ear, travels more slowly than light; it does not inform me whether the coming of the light is instantaneous or whether, although extremely rapid, it still occupies time." Then Salviati tells us that he has devised an experiment to determine whether or not the speed is infinite.

> Let each of two persons take a light contained in a lantern, or other receptacle, such that by the interposition of the hand, the one can cut off or admit the light to the vision of the other. Next let them stand opposite each other at a distance of a few cubits and practice until they acquire such skill in uncovering and occulting their lights that the instant one sees the light of his companion he will uncover his own. After a few trials the response will be so prompt that without sensible error the uncovering of one light is immediately followed by the uncovering of the other, so that as soon as one exposes his light he will instantly see that of the other.

From a modern viewpoint, his experiment seems silly. Silly, because we now know that over a three-mile range the light would take only about three-millionths of a second to travel the six miles back and forth.

It wasn't an experiment that determined the answer; rather, it was an accidental observation of the eclipse of one of Jupiter's moons. In 1676, Ole Roemer, a Danish astronomer, decided to investigate a mystery behind the eclipses of one of Jupiter's moons. He noticed a twenty-two-minute difference between the times of eclipses of Io when the earth was closest to and farthest from Jupiter (a distance of almost 300 million kilometers) and concluded that this difference was due to the extra distance that the light had to travel to earth. (Eclipses of Jupiter's moons should occur at regular intervals, since its moons travel at constant speeds.) He used this observation to make the easy calculation that the speed of light was about 225,000 kilometers per second. That was roughly 75,000 kilometers per second too slow by modern, established standards, but it did finally give a rather convincing argument that the speed of light is finite.

But why should the light spread out when passed through a prism? Why does the blue light bend more than the red? If light consisted of tiny particles, as Newton suggested, then how could beams of light pass through each other without colliding? Moreover, how could those tiny particles of light move from here to there instantaneously?

Roemer's argument should have at least convinced everyone that the speed of light is finite, especially since he was able to use it to predict lag-times of other eclipses. Christiaan Huygens, the seventeenth-century physicist and astronomer who first patented the pendulum clock, and Newton were convinced, but many others were not. They wanted a laboratory experiment, not a measurement depending on astronomical observations. After all, the French physicist Mersenne was able to determine the speed of sound in a laboratory.

Now we must fast-forward from Newton almost two hundred years to 1849 and the laboratory of Armand Hippolyte Louis Fizeau, who, working on a remarkably simple instrument, would not only demonstrate that the speed of light is finite, but also determine it to be 312,480 kilometers per second, surprisingly close to the most accurate speed known today (299,784 ± 10 kilometers per second). Fizeau's experiment was brilliant and simple. Fizeau's apparatus consisted of several lenses and mirrors, but the principal part was a rotating wheel with 720 teeth. Sunlight was focused through the top tip of the rotating wheel on a hill in Suresnes, France, and reflected directly back from a mirror 8.68 kilometers away in Montmartre in Paris. The teeth hacked the beam into stroboscopic flashes of light. When the wheel was at rest, and adjusted so the beam could pass between adjacent teeth, the light would be seen by the observer in Suresnes. But, at certain speeds, the light would be occluded to diminish its intensity. At other speeds the light would pass through one space between the teeth on its way to Paris and return through the next space. Fizeau carefully increased the rate at which the wheel turned until he observed the light at its highest intensity. At the best intensity, the light would have to make the round trip in the time it would take for the wheel to rotate from one tooth-opening to the next. This gave a clear measurement of the speed of light. He observed the highest intensity when the wheel was rotating at twenty-five revolutions per second. Therefore, the time between spaces was $1/25 \times 1/720 = 1/18,000$ seconds. The round-trip distance was 17.36 kilometers. Therefore, the speed of light would be $18,000 \times 17.36 = 312,480$ kilometers per second.

Fizeau's experiment convinced scientists that the speed of light is finite. Still, there were those who did not trust that Fizeau's experiment gave an accurate measurement of the speed. True, it was a laboratory experiment, but one that relied on Fizeau's judgment of the speed at which highest intensity is achieved. Besides, it seemed highly unusual for a simple laboratory instrument to measure speeds on the order of hundreds of thousands of kilometers per second with any accuracy. What was needed was an instrument that could measure the speed of light without using human judgment.

Jean Foucault built such an instrument. Beam a light source to a rotating mirror. The beam reflects and travels a distance 65.6 feet back to a stationary mirror, which, in turn, directly reflects the beam back again. But by the time the beam returns, the rotating mirror has moved to a new position, having moved through an angle α. The beam is reflected once again and projected at an angle 2α from the original source. Knowing the rotational speed of the mirror, Foucalt calculated the speed of light to be 298,000 kilometers per second.

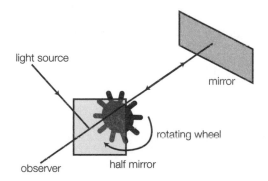

Albert Michelson perfected this experiment in 1879, and found the speed of light in a vacuum to be 299,853 kilometers per second. He also showed that all wavelengths of light in a vacuum travel at the same speed.

With light's speed measured so accurately, distance could also be measured with great accuracy. A pulse of light sent out in a certain direction until it strikes an object and is reflected back can be timed. If the speed of light is constant and accurately known, then the distance traveled is simply the speed of light multiplied by the time it takes to travel that distance.

But how does light move so fast? Unlike sound, it seemed to have no difficulty traveling great distances from galaxies billions of light-years away. Like Descartes before him, Newton believed that light traveled in small packages from its source, each far too small to be measured, though he certainly entertained the idea that light may be propagated as a wave. By that understanding he could explain how light travels in a straight line and how it reflects, and even how it refracts when it passes through different media. Light, for him, moved like billiard balls, bouncing off surfaces too dense to penetrate, passing through penetrable solids, and bending after initial interference. But he could not explain how light could pass through itself or how one beam could pass through another without resorting to the wavelike nature of light.

Huygens worked with the idea of *ether waves* back in 1656 to explain the phenomenon of one beam passing through another, but he could not explain how light could travel as a wave through nothing, nor could he explain the clear observation of light traveling in a straight line. *Ether* was an ancient Greek term used to explain the motions of heavenly bodies. Aristotle used it as the fifth element surrounding air. By Newton's time it was reborn to explain the interference of light—when light was considered to be a wave that needed a medium to propagate in. Ether permeated everything and pervaded all space, including the space between the atoms of solids. Its substance was a matter of debate and argument. Light casts sharp shadows because it does not seem to be able to go around objects. Sound, which was well known to be a wave, easily travels around objects in its path and does not cast sharp silences when blocked by an object. Its waves need a material medium to travel; it cannot travel through a vacuum. But light is different. It seems to be perfectly content traveling through a vacuum; in fact, it travels faster and more easily through a vacuum than through a material medium.

However, ether poses another problem. Light waves are transverse; they perform wavelike oscillations at right angles to the direction in which they propagate, like the waves that appear on a string being moved up and down, or ocean waves that simply rise and fall. Water waves never penetrate the water; they simply force the surface to undulate. (Sound waves are different; they are longitudinal, that is, they perform wavelike oscillations parallel to the direction in which they propagate, and can penetrate almost any material medium.) If light travels as a wave, it must be a wave of a material that is so rigid and pure that light could pass through it frictionlessly. Gases and liquids are too free to transmit transversal waves. What could ether be made of? Didn't light have to move through something? The answer was not what anyone expected. It was the next step in our understanding of motion.

THIS WAS THE age of Romanticism. The Napoleonic Wars were in full swing. Goya was painting his two *Majas*. Lord Byron was writing his first poems. Haydn was completing his *Seasons* oratorio, and Beethoven his *Eroica* symphony. Great, grand gestures were all around. In 1801, the English physicist Thomas Young, whose interests and accomplishments ranged from archaeology and philology to mathematics and medicine, performed an experiment to determine whether light traveled as a particle or a wave.

He projected a beam of light through two closely spaced holes onto a screen and observed that the beams split and recombined, producing a pattern of light and dark bands. He concluded that this could only have happened if the light was behaving as waves, with their crests or troughs reinforcing or canceling one another. Newton had claimed that light was corpuscular; his fame and stature were so great that his word was established as right. So Young's work was not accepted until the French scientist Auguste Fresnel was able to confirm that light was indeed wavelike in nature. But wave motion needs a medium in which to propagate, and so it was declared that there must be some invisible material medium filling the entire universe.

Some called it "luminiferous ether," suggesting that it permeates all matter, visible and invisible—vacuum, gases, glass, and any other material that light could penetrate. On the one hand, the ether seemed absurd; after all, such a rigid solid defied all experience. On the other, it was useful to justify not only the propagation of light, but also the notion of action-at-a-distance, which had been a concern ever since the force of gravity emerged as the reason for the movement of planets. Gravity may have been mathematically modeled as lines of force, but the question of *how* it worked was not answered. Was there stuff that gravity waves existed in? The early Victorian era was still mostly a mechanical world in which practical machinery used gears, pulleys, and levers to perform work. Educated people had not seen action-at-a-distance other than that performed by God or magnets. They may have known how Newton's gravitation laws explained paths of planets, and the more informed physicist knew about gravitational force fields (lines of force between masses), but how, exactly, does the physical world conform to those mathematical fields? Orbiting planets are not tethered to the sun by visible ropes. And how did magnets work anyway?

In 1820, the Danish physicist Hans Oersted connected a battery (which had been invented in 1800 by the Italian physicist Alessandro Volta) to a wire loop and watched a compass needle move when placed near the wire. The "flow" of electricity—the *electric current*—had induced its own magnetic force, a force that would happen only when the wire was connected to both negative and positive poles of the battery. Then, in 1831, Michael Faraday, the British Victorian experimentalist, generated electricity by moving a magnet around a wire (the principle of a dynamo), decisively linking electricity and magnetism. Electricity was previously thought of as some sort of fluid that flowed within the wire. Faraday hypothesized that there must be some physical phenomenon within the empty space between magnet and wire that would enable the magnetic force to travel through space to cause the flow of electricity. He explained the physical phenomenon as a field of invisible lines surrounding the magnet that describe the magnitude and direction of the force.

Mainstream physicists did not take Faraday's field idea seriously, partly because it seemed crazy to believe that a field of ghostly lines of no tangible substance could have physical powers, and partly because established theoretical physicists slighted his work for its lack of mathematical language. His theory was viewed as nonsense.

The world was grasping for an understanding of what

things, especially light, but planets, too, moved through. Could fields provide forces without physical things to push or pull? The field is a space of points. At each point there is a description of what will happen to an object positioned or passing through that point. That description may include a direction to follow at a certain speed, a force to cope with, or anything else to inform behavior and change in behavior. For example, each iron filing in a magnetic field is told to orient itself in a certain direction according to its place in the field. In the case of a gravitational field, each mass is told to accelerate to a new position where it will be given new instructions on where to go from there.

In 1864, the same year that the first Sholes & Glidden Type Writer was being assembled in Ilion, New York, the Scotsman James Clerk Maxwell published a monumental treatise on electricity and magnetism built on the work of Faraday, in which he claimed, "light consists in the transverse undulations of the same medium which is the cause of electric and magnetic phenomena." Suddenly, the speed of light became an extremely important number for physics. Maxwell's equations showed that electricity and magnetism were inseparable components of something more comprehensive. It was as though they were silhouettes of each other on two sides of a screen. Electricity passing through a wire produces a magnetic field. A magnet moving around a wire produces electricity. Maxwell neatly modeled all this with just four interrelated equations, providing a simple account for action-at-a-distance through invisible lines of force.

Maxwell had given a mathematical description of Faraday's field idea. Maxwell's equations showed that there is a wave effect—visualized simplistically as the sort of wave one sees along a string that is fixed at one end and jiggled at the other—that transmits electric and magnetic forces from an energy source, point-to-point, along Faraday's field lines. Any disturbance of an electric or magnetic force is transmitted through the field like a wave along a string. Moreover, a changing magnetic force induces a changing electric force and vice versa. The symmetry of this interaction generates a sympathetic leapfrogging of three-dimensional waves—waves of electricity and magnetism perpendicular to each other.

Changing the electric field induces a change in the magnetic field, which induces a change in the electric field . . . on and on this system goes, supporting a dynamic and inseparable marriage of electric and magnetic waves, an effect that could be mathematically modeled as a three-dimensional *electromagnetic* wave. Maxwell's equations implied that magnetic and electric waves are inseparable components of more comprehensive electromagnetic waves.

Had Maxwell's description come as a purely mathematical deduction, it would have been a matter of theoretical consequence, a conjecture, not physical proof of the existence of electromagnetic waves. It remained a conjecture for a quarter century—under a great deal of popular opposition—until the German physicist Heinrich Hertz established the physical evidence of electromagnetic waves by detecting radio waves, which were predicted by Maxwell's theory.

It seemed as though the last piece of the 2,000 year puzzle of action-at-a-distance was in place. The electromagnetic field could not be seen, but it could radiate as a wave and have its effect anywhere in space as far as the wave could travel. When Maxwell calculated the speed of electromagnetic waves he was surprised to find that it was the same as the speed of light. Could it be a coincidence? Could light, itself, be one form of electromagnetic radiation? Perhaps what we call light is electromagnetic radiation of a certain wavelength. Perhaps we see only a small portion of the whole spectrum of electromagnetic waves. We hear only a small part of the sound spectrum, so perhaps there are wavelengths of radiation that the human eye cannot absorb because they are too short or too long.

Question the continuity of light as an electromagnetic wave and we question the continuity of what is waving. Is the space-filled ether continuous? Zeno's paradoxes of motion were transposed into questions of electromagnetic radiation, not just visible light.

The electromagnetic field was introduced as a physical entity distributed throughout space, pulling and repelling according to its charge by an inverse-square law similar to that governing the behavior of gravity. The force field could only be measured by observing the acceleration of test bodies placed within the field.

Maxwell's equations verified Christiaan Huygens's hypothesis that light is a wave. Out of this came a generalization of Newton's laws that described the relationships and interactions between charged particles as well as electric and magnetic fields.

Moreover, Maxwell had achieved what Descartes long ago claimed should be the focus of science, that the nature of reality should be entirely expressible through a small number of general laws. With just four simple equations (two pairs of symmetric equations), he was able to express the continuity of electric and magnetic fields and how changes in one field effect changes in the other.

On the tail of Maxwell's discovery came two extraordinary adjustments to classical physics. One was relativity, the other was quantum theory. The more scientists looked at how things moved, the more mysterious the medium that things moved through became. Calculus had mathematically wed time to motion. Maxwell began a new mathematical courtship with space. Throwing off convention, Einstein married them all.

12

The Space-time Revolution

Close to a decade before Einstein published his famous papers on special relativity in 1905, a young man set off on an ivory, nickel, and crystalline quartz machine to travel 80,000 years into the future, slipping "like a vapor through the interstices of intervening substances . . . out of all possible dimensions—into the Unknown."

Exactly one week later he returned, looking hardly any older than he had when he vanished from his laboratory in a milky gust of whirling air. The room was just as he had left it, his tools and equipment unmoved. Trembling violently, and shaky on his lanky legs, he brushed dust and dirt from his tattered evening clothes. Soon afterward, his guests found him smoking his pipe against a lilac-scented breeze coming from tall French windows open to a garden, eager to tell about his most implausible adventure.

His trip, which must have lasted a good deal more than an earth year, gave him a disturbing glance of the descent of the human race. Yet on his return, just one earth week later, both he and his guests seemed to have aged equally.

H.G. Wells wrote *The Time Machine* in 1895. He could not have known—yet—that time is not absolute; that it is intimately connected with space; and that it depends on a frame of reference. He was aware of Zeno's claim that "motion is an illusion," and accepted Descartes's beliefs that the external world of matter and motion is known only by the senses; that there are no colors, no smells, and no textures other than what we make of them; that all our sensations are illusions. Einstein was about to announce, "The distinction between past, present, and future is only an illusion."

Time and space had been considered disconnected ingredients of physics. Speed, measured in distance per second, gave a mathematical linkage between space and time. An object could move through space, but moving through time would have been nonsense before Einstein's theory of relativity.

THROUGH THE NINETEENTH century, scientists presumed that light traveled in ether. This ether was identified with space, assumed to be continuous, thought to be the conduit responsible for action-at-a-distance, and was the reference point against which all motion took place. This was the belief before two American scientists set out to measure the ether.

When Leonard Case, Jr., a rich lawyer, died in 1880, his family's home was, by bequest, turned into the Case School of Applied Science. It was a plain beige stone mansion with a handsome old barn on a wide sycamore-lined street near a park and public square in the center of Cleveland, Ohio. A year later, Albert

Michelson, who had just turned twenty-nine, arrived as the school's first professor. He had no degrees of higher education, although he had graduated from—and later taught at—the U.S. Naval Academy. He wasted no time in setting up chemistry and physics laboratories on the second floor of the barn. There were sixteen students at the school, and tuition was $100.

Across the road on the north side of the tracks for the New York, Chicago and St. Louis Railroad was Western Reserve University, which was mainly housed in Adelbert Hall, a four-story buff limestone building with a tall, thin clock tower. Edward Morley was a self-trained chemist who had been professor of natural philosophy and chemistry at Western Reserve for thirteen years.

The two scientists could hardly have been more different. Michelson was fifteen years younger than Morley, a handsome man with deep, dark eyes and a trim sideburn beard reaching down to his chin, well groomed, and agnostic. Morley was the stereotypical preoccupied professor who dressed without care and whose uncombed hair went down to his shoulders. He was a deeply religious man who delivered sermons in the college chapel and local churches. His red moustache was so large that it covered his mouth and ran back to his ears. Yet the two scientists shared certain habits. Michelson played the violin and Morley the organ. Both were hands-on experimentalists with a passion for precise detail.

Michelson had built instruments to measure the ether drift, or motion. He was obsessed with an entry in the *Encyclopedia Britannica* written by James Maxwell: "If it were possible to determine the velocity of light by observing the time it takes to travel between one station and another on the earth's surface, we might by comparing the observed velocity in the opposite directions determine the velocity of the ether with respect to these terrestrial stations." Just before his death in 1879, Maxwell published a letter in the British scientific journal *Nature* doubting that anyone could determine the speed of light as he suggested. Michelson surely knew about Maxwell's letter, and it must have energized his fixation on measuring the speed of light.

When Michelson told Morley that he had a plan to test the effects of ether on the speed of light, Morley suggested that they work together at his lab in the basement of Adelbert Hall. Experiments to determine ether drift had been conducted since Fizeau and Foucault experimented with the speed of light in the 1850s. Michelson had made some modifications to Foucault's methods while at the Naval Academy, but he wasn't satisfied with the results. When Morley invited Michelson to work on the experiment in his lab, Michelson had no college degrees— not even a bachelor's degree. Morley had an international reputation for having found percentages of oxygen in air, the relative weights of oxygen and hydrogen in water, and the weights of a liter of oxygen and a liter of hydrogen.

It was 1887 when Michelson and Morley set out to build an instrument to measure the effects of ether on light. An extraordinarily delicate and massive apparatus was required—the slightest vibration would render the research useless. A square stone fourteen inches thick floated on a pool of purified liquid mercury to minimize any natural vibrations coming from tremors in the earth or the building, and also so that the stone could be easily, yet slowly, rotated. (One can imagine the poisonous vapors in that laboratory.) Supporting columns rested on the bedrock below the building.

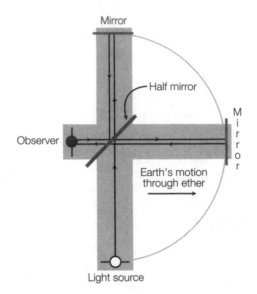

Two perpendicular bars held mirrors at their ends. A beam of light was sent along one bar to a half-silvered mirror, which split the beam, allowing half to pass directly through and half to be reflected at a right angle along the length of the other bar. A mirror at the end of each bar reflected each half-beam back to an observer. If the earth were moving through the ether, the wavelength of light in the direction of the earth's movement should have been shrinking by having to move against the movement of the ether. The observer would then have seen the two returning beams out of phase. The returning light would be darkened by the one wave interfering with the other. The floating instrument could be rotated to find the earth's most pronounced movement through the ether.

But Michelson and Morley were astonished to find that the two rays returned at precisely the same time, perfectly in phase. Rotating the perpendicular rays through different angles made no difference. It became known as "the greatest of all negative results." The velocity of light did not depend on the direction of motion or the speed at which the observer was moving. No matter how fast the light source moves, the light's speed is the same. They knew that the earth travels in its orbit at a speed of about twenty-nine kilometers per second. They expected to find that light traveling in the direction of the earth's movement around the sun should be greater than that in the opposite direction. This was very strange.

Michelson suggested that "the earth drags the ether along at nearly its full speed, so that the relative velocity between the ether and the earth at the surface is zero or very small." All along he had assumed what all scientists up to that point had assumed—the existence of ether. An experiment showing that traveling light waves were unaffected by ether would be considered a failed experiment. But he persisted, "Since the result of the original experiment was negative, the problem is still demanding a solution."

What could possibly be the explanation? Perhaps the earth itself is being dragged by the ether and is moving like a cork floating in water along with the ether. To test that, Michelson and Morley waited six months. If the earth were "floating" in the ether, it would have to be going in the opposite direction relative to the ether. But again, the two rays returned at the same time. After the apparatus was swung around 90 degrees, the two rays returned at the same time. And after the instrument was turned in other directions, the two rays returned at the same time.

There are two reasonable explanations: (1) the simple mathematical calculation is wrong, or (2) there is no ether. Either explanation would have dealt a catastrophic blow to classical physics. Mathematics surely had to model reality to some extent, and the absence of ether—the sea of the universe, against which positions of all bodies could be marked and all motion could be detected—would have meant that there is no such thing as absolute motion. But a more serious blow would have been to the wave theory of light. How could a wave travel through no medium at all? What would be waving?

TWO YEARS AFTER the Michelson-Morley experiment, the Irish mathematical physicist George Francis FitzGerald, referring to the experiment, suggested that perhaps the distances that each light ray travels are not equal. Perhaps distance measured along the stream of the ether shrinks in the direction of motion at speeds close to that of light. Perhaps the measuring instrument itself depends on the direction of measurement.

Here's the rough idea behind FitzGerald's explanation. If you send out a ray of light in the direction of the earth's travel through the ether and reflect it back by a mirror fixed at a certain distance d, then it will travel from its source to the mirror at velocity $c - v$, where c is the velocity of light and v is the velocity of the ether. We subtract the two velocities because the light

would have a headwind caused by the mover
ether. But on return (after reflection) its veloc
because of the tailwind. The time it takes t

trip would be $y = \dfrac{2dc}{c^2 - v^2}$. This is simply because time is equa.

to distance divided by velocity.

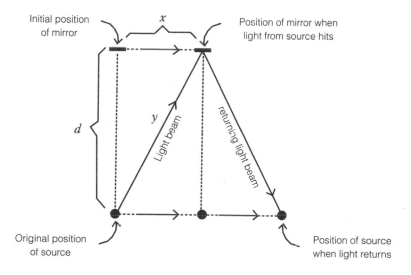

Now let us play the same game with a mirror located at the
same distance in a direction at right angles to the ether. There is
no tail- or headwind. Let d denote the distance from the source
to the mirror. (See the illustration above.)

While the light travels, both source and mirror have been
moving in the direction of the ether (to the right in the illustration), and therefore by the time the light reaches the mirror,

,oth source and mirror will have moved to a new location x units away. The distance the light traveled, y, is related to d and x by the Pythagorean theorem, so: $y^2 = d^2 + x^2$.

Now here is the critical move. Both the light and the mirror arrive at the mirror's new position simultaneously. That means that the light must move through a distance y at velocity c and the mirror must move through a distance x at velocity v. Therefore, $y/c = x/v$ and hence $y \cdot v = x \cdot c$. By the Pythagorean theorem applied to the upper left triangle in the illustration, $y^2 = d^2 + x^2$ is the square of the distance the light must travel to catch the mirror. Putting these last two equations together, we find that $y^2 = d^2 + \left(\dfrac{yv}{c}\right)^2$. Solving this for y gives $y = \dfrac{dc}{\sqrt{c^2 - v^2}}$. This is the distance the light travels in order to reach the mirror. It travels the same distance for the return trip. So the total distance it must travel is $y = \dfrac{2dc}{\sqrt{c^2 - v^2}}$.

So how long does that round trip perpendicular to the source's motion take? The answer is to divide the distance by the velocity of light c. It takes $\dfrac{2d}{\sqrt{c^2 - v^2}}$ units of time. Let us compare the time it takes in the direction of the source's motion to the direction perpendicular to the source's motion by dividing one with the other:

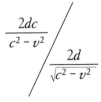

$$\dfrac{\dfrac{2dc}{c^2 - v^2}}{\dfrac{2d}{\sqrt{c^2 - v^2}}}$$

With a little algebra this last ratio simplifies to

$$\frac{1}{\sqrt{1 - \frac{v^2}{c^2}}}.$$

This is the ratio of the time it takes to travel in the direction of the source's motion as opposed to the direction perpendicular to the source's. This number is always greater than 1 whenever v is greater than zero. It means that *as long as there is ether, it will take longer for the light to travel in the direction of the source's motion than to travel perpendicular to the source's motion.* If there is no ether, or if the source is traveling at exactly the same velocity as the ether, then $v = 0$ and the ratio would be 1.

FitzGerald suggested that matter shrinks in the direction of motion, shedding light on why Michelson and Morley's experiment could not detect ether. The idea is simple, albeit strange: The apparatus shrunk in the direction of the earth's motion by an amount precisely equal to offset the differences in distance. This is the wonder of absolute motion: It conspires with our measuring instruments to prevent any possibility of detection.

FitzGerald figured that the shrinking happened in the direction of motion and increased with the speed of the object. With slow-moving objects such as trains, the shrinking would be negligible. But at fast speeds, such as the motion of atomic particles, the shrinking would be considerable. To a stationary observer, a twelve-inch ruler traveling in a vacuum at half the speed of light in the direction of its length would appear to be less than ten inches long. The faster it moves the shorter it gets. At the speed of light it disappears entirely—it has zero length. A logical deduction from this is that nothing can move faster than the speed of light in a vacuum, for otherwise the moving object would have a negative length.

The Dutch physicist Hendrik Lorentz worked out a similar formula relating the mass of a charged particle at rest with the mass when it is in motion. A ruler with a mass of 12 grams traveling at half the speed of light would increase its mass to approximately 13.8 grams. At the speed of light its mass would be infinite. Lorentz's mass increase is similar to FitzGerald's shrinking. In fact the percentage of Lorentz's mass increase is the same as the percentage of FitzGerald's shrinking.

Here, mathematics is making a prediction that contradicts our common-sense intuition about the physical world. Could the ideas of FitzGerald and Lorentz be demonstrated—even indirectly—by some measurement? An electron's charge does not increase with velocity, and its mass/charge ratio can be measured by a deflection of the electron in a magnetic field. An increase in mass/charge ratio with velocity would imply that the mass increases with velocity; this would give indirect proof of Lorentz's hypothesis. In Berlin in 1897, while investigating the deflection of electrons in electric and magnetic fields in vacuum tubes, the German physicist Walter Kaufmann discovered that the mass/charge ratio did increase with velocity.

EINSTEIN'S FIRST CONCEPTION of relativity in 1905, special relativity, proposed that nothing is faster than a photon, a particle of light. What about gravity? How fast does it travel? How quickly would the earth sense a mass that just happened to come into existence a billion miles away? Surely, this new existence would not be felt instantaneously on earth—that would mean that something must have traveled a billion miles at a speed faster than a photon to give us the information that the mass came into existence, contradicting special relativity.

Newton's universal gravitation equations provided an excellent mathematical model for gravity that could perform excellent predictions, but it did not explain how gravity works, nor tell us how it travels, nor tell us what it is.

Since Newton, gravity had gotten weirder and weirder. According to special relativity, we should not be able to sense motion while traveling at a constant speed in a windowless train because motion does not make sense, unless we are observing our movement relative to something else. In fact, all the laws of motion (if not all the laws of physics) would appear to be the same to anyone traveling on that windowless train. But we do sense movement when we accelerate, even if we are in a windowless train, as everyone who has ever been in a fast accelerating elevator knows. We feel that acceleration as gravity. The principal idea here is that gravity and acceleration are so firmly linked to each other that motion itself may be used to understand the phenomenon of gravity.

The Newtonian explanation of why the moon stays in its orbit around the earth involves gravity, the mutual force between earth and moon; the earth wins in this continuous tug-of-war and pulls the moon down as the moon tries to continue in motion along a straight line. The net effect is a nearly circular orbit of the moon about the earth, as though the moon were tethered to the earth. But what is the stuff of that terribly long, invisible tether?

The theory of general relativity, which Einstein published in 1915, has an answer using an intuitively simple geometry. We make a slight simplification by imagining space to be two-dimensional rather than three. This simplification is far more visual than its three-dimensional analog, because it frees up one dimension so we can imagine ourselves looking at space from one dimension higher, the way you are looking at this two-dimensional page from outside this book.

So imagine space as an infinite two-dimensional flat surface that we are looking at from a third dimension. Further, imagine that the material of the space is nylon and that it stretches taut like a trampoline, but that a dimple appears wherever a mass is placed on it. The dimples in the three-dimensional analog more accurately represent gravity in our relativistic universe. They are just harder to visualize in three dimensions.

Each dimple represents a mass—the larger the mass, the larger the dimple. A ball will move on this trampoline with a velocity and acceleration that is determined by its proximity to a nearby dimple. (See illustration below.)

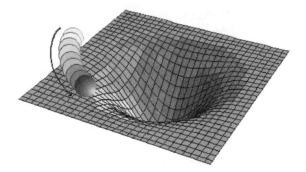

A ball rolling far from any dimple would follow a straight Euclidean line because it would be rolling on a reasonably flat portion of our model. And, of course, the mechanics of gravity would work just as Newton said they would, just as though the mass was a rolling ball near a dimple on a nylon trampoline. However, the trampoline analogy only goes so far as to say that the space around masses is warped by gravity, so we should be careful to understand that rolling masses near dimples don't actually fall down into the dimples. For general relativity, space is curved sharper near masses and flatter where there are none.

This new geometry of space offers an understanding of how matter moving in space exerts force. The mysterious invisible tether holding the moon in orbit about the earth is the stuff of dimpled space geometry. More massive material causes deeper and larger dimples. Moving masses create moving dimples. This picture of the universe is a whirlwind of moving dimples, some floating and orbiting as satellites of stronger (deeper) dimples, others being absorbed by subordinate neighbors like soap bubbles wandering through space swallowing one another.

Dimple geometry is the geometry of general relativity, a geometry in which a straight line means the path of shortest distance. In Euclidean geometry, the shortest distance between two lines is the usual Euclidean straight line. But general relativity insists that the universe is not Euclidean, that even if a straight line appears straight, it is only straight in the sense that its distance on our model trampoline is the shortest path from where it begins to where it ends. Moreover, we begin to see how the force of attraction comes into play—the dimple's incline increases as it gets closer to its center.

If the moon were traveling in a Euclidean straight line and the earth were to suddenly appear as a dimple, the earth would make ripples in the fabric of the universe, just as a bowling ball would if it were to suddenly appear on a nylon trampoline.

Those ripples would radiate from the center of the space occupied by the center of the earth and dissipate as they receded to outer space at the speed of light—it seems that gravity does not instantly affect masses billions of miles away.

In this geometry, the universe is relatively bumpy with dimples wherever there is a mass, but the bumps are more or less smooth, even near heavy masses. This smoothness—which in calculus terms means infinitely differentiable—requires continuity.

What goes for space goes for time. If gravity is another form of acceleration, then acceleration—hence time—must be joined to space at the hip and both are bent by gravity. This is the essence of the space-time revolution. Acceleration implies a time component, so there is one more dimension that was ignored in our dimpled trampoline analogy. But how do we see time as a special dimension?

First we must understand what we mean by dimension. To the mathematician, dimension is analogous with our real-world spatial dimension, a generalization of the word "dimension," which has ancient Greek roots ($\delta\iota\sigma\ \mu\varepsilon\tau\rho\rho\nu$) meaning *twice measure*. Greek geometers understood that the word has a more general meaning than specific measurements in two directions. The word conveys the notion that *twice measure* could be could be also mean *thrice measure*.

Science fiction writers exploit the notion that if only we could understand how to jump in and out of higher dimensions, we could go back and forth in time and play all sorts of magic tricks with our minds and bodies.

We use the word loosely as a reminder that it comes from measuring our visual surroundings—a curtain has such-and-such length and width, a couch has a length, width, and height—so we tend to want to visualize higher dimensions whenever we hear that mathematics uses some higher dimension to solve a problem, but they are really just an array of measurements (functions or numbers). Physicists use these arrays in equations to describe the world.

A point in space has an address (x,y,z) given by three real numbers, x, y, and z, as measured along three perpendicular lines, the x-, y-, or z-axes. If that point happens to be moving at some nonuniform speed along the x-axis, we could represent that speed by a fourth number, v_x, and tag it onto the other dimensions. We would have a point with not only an address, but some information about how fast that point is moving in the direction of the x-axis. Or, to keep track of events in space as time passes, we have the four-dimensional address (x,y,z,t), where t represents time.

Thus we live in four dimensions. But we should keep in mind that this is just a man-made scheme for understanding the world through mathematics. Dimension is a human-made word, not a God-given one.

The great advantage is that the notation (x,y,z,t) provides an easy way to keep a good record on the history and future of a moving point. Now, it may be that the speed v_x depends on where this point is in three-dimensional space. For example, v_x may equal $x + y + z$. If so, the point could be represented as $(x,y,z,x + y + z)$. There are four numbers, but the last depends on the other three and therefore only three numbers determine the position and speed of the moving point. If the point were moving through three-dimensional space, it would be moving on a three-dimensional surface in four-dimensional space.

According to the Russian-German mathematician Hermann Minkowski, the world is not three-dimensional as we have tended to think of it, but four-dimensional with points described by *(x,y,z,t)* in a so-called *space-time continuum*. It means that any event or movement that happens—be it an arrow moving through space or just standing still, or you reading this book—is represented by some graph in the space-time continuum. That graph (called the *worldline*) depicts the progress of the event as time passes; it may be a line, a curve, or an area. Space and time are expressions of a single quantity, space-time. "[With] the multiplicity of all thinkable *x,y,z,t* systems of values," he wrote, "we will christen the world." This last variable is time. If any one of the variables changes ever so slightly, the point in the fabric of space changes ever so slightly. But time is not fully independent of the three dimensions of space; according to Lorentz's transformations, it contracts along with space when observed from a relative frame of reference.

We can picture this in one dimension, say the x-coordinate. Our simplified space-time has two dimensions (x,t) and the Lorentz transformation of a measured time interval from one reference frame of a moving object to another is

$$\Delta t' = \frac{\Delta t - \frac{v \Delta x}{c^2}}{\sqrt{1 - \frac{v^2}{c^2}}},$$

where $\Delta t'$ is the observed time interval in one reference frame, v is the velocity of the object, Δx is change in the x-coordinate, c is the constant speed of light, and Δt is the observed time in the second reference frame. We gather from this that the time interval as measured in one reference frame is faster or slower than the other—it doesn't matter which. According to another Lorentz transformation, space also dilates when measured in reference to a moving object. We have

$$\Delta x' = \frac{\Delta x - v \Delta t}{\sqrt{1 - \frac{v^2}{c^2}}},$$

where Δx represents the change in position in one reference frame and $\Delta x'$ represents the change in position from the point of view of the observer. In other words, time is as warped as space. Now watch what happens when we manipulate these last transformations. By simple (but messy) algebra we arrive at the following:

$$c^2 \Delta t'^2 - \Delta x'^2 = c^2 \Delta t^2 - \Delta x^2.$$

What can we gather from this? Notice that the changes in variables on the left are those that would have been observed from one reference frame and those on the right from another. But the quantity on the right is equal to the quantity on the left. In other words, the observers in any particular reference frame will arrive at the same number (an invariant interval of space-time) when computing $c^2 \Delta t^2 - \Delta x$.

This last expression is denoted Δs^2. We have $\Delta s^2 = c^2 \Delta t^2 - \Delta x^2$. Any equation of the form $a^2 = x^2 - y^2$, where a is a constant, graphically represents a hyperbola. So $\Delta s^2 = c^2 \Delta t^2 - \Delta x^2$ represents a hyperbola in one-dimensional space-time.

This makes sense if we think of what happens when we spin a ball tethered to a string. The ball is under constant acceleration, and by the Lorentz transformation the ruler measuring the circumference of the circle from the ball's reference point is longer than what would be predicted from the Euclidean geometry (flat surface) calculation. The circle of Minkowski's space-time does not sit on a flat plane, but rather in hyperbolic space, where the normal calculations for Euclidean distance do not apply. For Euclidean geometry, distance is measured according to the Pythagorean theorem, which says that the square of the distance between points, Δs^2, equals the sum of the squares of the coordinate intervals, i.e., $\Delta s^2 = \Delta x^2 + \Delta y^2$. However, in Minkowski space-time, the square of the interval is not equal to the sum of the squares of the coordinate intervals but to the difference. Space-time is not Euclidean; rather it is hyperbolic.

Returning to four dimensions of space-time, we find a similar picture. An interval in space-time Δs is independent of the reference frame and can be measured by its four-dimensional coordinates as $\Delta s^2 = \Delta x^2 + \Delta y^2 + \Delta z^2 - c^2 \Delta t^2$, where Δs^2, Δy, and Δz represent the position differences between the two events and Δt the time difference. (Note how similar this is to the Pythagorean theorem that is applied in Euclidean space when $\Delta t = 0$; i.e., the distance between two points whose coordinate differences in Euclidean space are Δz, Δy, and Δz, and is given by $\sqrt{\Delta x^2 + \Delta y^2 + \Delta z^2}$.)

General relativity, the theory applied to the viewing space on the astronomical scale, links space and time in a four-dimensional mathematical curved universe that geometrically models the mass of an object as something that depends on its velocity, its rest mass (mass at zero velocity), and the velocity of light. The relativistic mass of this object, though dependent on its velocity, still has the property that the force acting on it is equal to its mass times its acceleration.

Einstein's celebrity has mythologized how such a brilliant mind could not get a researching job, even as an assistant, at a respectable academic institution and be called "a lazy dog who never bothered about mathematics at all" by his teacher Hermann Minkowski. But it is true. Even when he was in elementary school in Munich, his teachers said that he had no future in science. All through his elementary and gymnasium education he was unhappy and did not excel in school. At fifteen he was expelled for disruptive behavior and disrespecting his teachers.

Evidently school is not the only source of intellectual stimulation. As with many successful people who had disagreeable schooling, all it took was one loving mentor to fire a spark. Albert's mentor was his uncle Jacob, an electrical engineer who spent endless hours talking science and electricity with Albert. Albert had failed the entrance examination to the Swiss Federal Institute of Technology and had to spend a year getting tutoring help at a small country school instead. The next year he was admitted to the prestigious school, from which, in 1900, he received his diploma but not a Ph.D.

By 1902, he still had insufficient references from his teachers for a good academic job. To earn a living, he took on various meager, short-term jobs before accepting a position as a patent officer at the Swiss Patent Office. While at his desk at the patent office in Bern, he worked on scientific problems, presumably in his spare time. In 1905 he still did not have a Ph.D., but he published three papers that rocked the foundations of Newtonian science. Through very elementary mathematics, he was able to

establish the shocking revelation that time itself—like FitzGerald's length and Loretz's mass—is relative. Newton's universal
time was dead. Things moved in a way we had previously
never imagined.

In a letter to the English philosopher Herbert Samuel, Einstein wrote that physical reality comes through our consciousness and complexes of sensations. He wrote about the table in
his room, suggesting that it was "merely a complex of sensa-
tion" to which he assigns a concept and name: ". . . [O]ne is in
danger of being misled by the illusion that the 'real' of our daily
experience 'exists really,' and that certain concepts of physics are
'mere ideas' separated from the 'real' by an unbridgeable gulf. In
fact, however, positing the 'real' that exists independently of my
sensations is the result of intellectual construction. We happen
to put more trust in these constructions than in the interpretations which we are making with reference to our sensations."

The four-dimensional space-time continuum tells us a great
deal about space as viewed from cosmic distances. It is smooth
in the sense that—even at its bumpiest spots—it has tangents.
Aside from things called black holes, which are massive spikes
in the nylon, the dimples for huge masses are still fairly smooth
when looked at closely. But general relativity is farsighted and
cannot see the microscopic texture of space. For that we must
go to quantum physics. And that is not an easy transition. Einstein did it, but it baffled him—and just about everyone else
since.

· 13 ·

Oops, Things Get Grainy Again

 At the turn of the twentieth century, the earth was partitioned by European imperialism. The continent had been at peace for the relatively long period of twenty-two years, though wars smoldered around the globe.

It was Germany's time "for a place in the sun." She was producing more steel than France and Britain, had a progressive chemical and manufacturing industry that outmoded Britain, and was by far the most educated country in the world, with the then-unusual system of government and industry supporting basic research without regard to profitable application. A strong educational system generated vigorous scientific research, which young energetic students were applying directly to industry. British manufacturers, fearing "science [going] abroad like an unloved child," cried foul. They felt it was wrong for a country to support ". . . a nasty class of intellectuals to interfere in the affairs of sound business men."

In Berlin, high, vaulted ceilings and gilt-columned interiors adorned tea salons, just as they did at grandiose concert halls built to symbolize the epoch and demonstrate civic pride, where Austro-German musicians—Mahler, Brahms, Strauss—regularly conducted their own masterworks. Each morning, crates of orchids from Hanover greenhouses were brought by train to be placed over white linen tablecloths at the teahouses along the Leipzigerstrasse, where women in bright colors and men in three-piece suits would meet for strudel and coffee. Women wore enormous hats, and though respectable women would not show their ankles in public, their elaborate gowns teasingly exposed their bosoms. New ships with their own concert halls were built to cross the Atlantic from Bremerhaven; trains were carrying guns as big as houses; mansions and apartments with fourteen-foot-high ceilings were built in the Tiergarten; and the Berlin middle class was becoming so large it was almost the majority.

The electron had recently been discovered and identified as a particle whose mass is on the order of only a thousandth that of the hydrogen atom, but very little was known about the structure of even the simplest atom. The scientific world was still recovering from the shock that even the atom has parts.

ZENO WANTED US to consider positions infinitely close to each other and to explain why his arrows move from position to position, though he could not have known that even mathematics cannot tell one position in space from the next, since there is no next. With calculus, mathematics described how positions that are infinitely close to each other pass along the act of motion by giving a unique state of position and velocity whenever the time key is pressed.

Aristotle thought matter was consistent, all the way down to the "uncuttable" atoms; a piece of wood is wood all the way down, beyond sawdust, to its most elementary parts. At the end of the sixteenth century the notion of atomism was revived with two new assumptions: that reality might not be homogeneous all the way down as Aristotle had assumed, and that the true description of nature might not be delivered by our senses. It now seemed that the mathematics appropriate for describing the observable experience level of the laboratory *had* to be different from the mathematics needed to describe the atomic level.

In 1897 the British physicist J. J. Thomson discovered the first subatomic particle, the electron—the lightest known particle having a nonzero rest mass—and in 1919 the New Zealand nuclear physicist Ernest Rutherford discovered the proton. By the time James Chadwick discovered the neutron in 1932, only four elementary particles—those pieces of the atom not made up of smaller particles—were known. Today there are hundreds of known elementary particles coming from nuclear accelerators and the cosmic rays of outer space—some, the leptons, are less than 10^{-19} centimeters in diameter.

Quantum physics, the theory that enables us to look deeply into the subatomic world, does not join easily with general relativity. If the world were as orderly as science expects it to be, there would be a single theory compatible with both general relativity and quantum theory, a model that can work as both a telescope to sharply collapse vast distances and a microscope to magnify subatomic scales. It would be a model that depended on a continuous sliding variable that could shrink vast distances or expand tiny ones.

Post-Newtonian mechanics was thought to have addressed Zeno's concerns. The infinitesimal calculus established a reasonable description of motion as a continuous function of position with respect to time. Every possible state of motion has been examined by those continuous functions and their derivatives so that any instant can be inspected for a position, velocity, and acceleration. Moreover, calculus has provided us with soothsaying differential equations, which in most circumstances can, for any specified future time, unequivocally predict precisely where the object will be. For centuries matter was thought of as being localized in space, with an address that could be determined by specific coordinates and a character described by its energy and momentum. We once thought that if a particle's position, energy, and momentum were known at some initial time, then its new position, energy, and momentum could be determined for some future moment. Quantum mechanics said no.

In 1927 Werner Heisenberg was working at the University of Göttingen, in a small charming walled town on the River Leine sandwiched between rolling mountains and famous for its lime trees, sausages, beer, and influential mathematics. Göttingen was still one of the great European centers of science and mathematics when Heisenberg discovered his *uncertainty principle*.

Every observation disturbs the very measurement that is undertaken. Even if we could build the ultimate measuring instrument, it would still be impossible to simultaneously measure with absolute accuracy both the position and momentum of an electron. Every attempt to measure both position and momentum will give errors in both, the product of which must be less than Planck's constant *h*. We are led to conclude that particles cannot be detected at instantaneous velocities or positions, but only judged to be in motion over sequences of discontinuous intervals without any possible observation of what goes on between those intervals. All we can possibly know is that the particle has a probability of being within some region of space and that its energy and momentum also have some probability of being within certain measurable bounds.

Our observations and investigations will always have at least this degree of indeterminateness, but quantum mechanics characterizes that indeterminateness by a mathematical wave that (when squared) gives the probability that our particle is within a region of space and has energy and momentum within a particular range. A particle has a mathematical representation of a wave associated with it—a wave function—that permits the approximate agreement between observation and theory. It is a wave that tells about the motion of the particle by giving a probability that it can be somewhere at a particular position in space and time.

The wave is not an actual wave that the particle moves along. It is a continuous function that predicts probable positions for any future instant if some aspect of an earlier position is known. However, the probabilities given are just that, not certainties. Indeterminate states of motion are built into atomic behavior. But the uncertainties at the atomic level are so small that their accumulation at the human level is still within the bounds of any experimental error attributed to instrument in-accuracies.

As Louis de Broglie, one of the twentieth-century founders of quantum mechanics put it, ". . . the essential indeterminate-ness is completely masked by the errors introduced in the course of experiment, and everything happens therefore as though it did not exist at all. In other words, each corpuscle at each of its manifestations has always, so to speak, to make the choice between several possibilities; but the limits of this choice are supposed to be so narrow that in practice as also in experi-ment, everything happens as though instead of free choice there were a strict Determinism."

The quantum mechanics story began when a German physicist named Max Karl Ernst Ludwig Planck asked why subatomic particles radiate a blue light when they travel through a non-vacuous medium faster than the speed of light in that medium. When iron is heated it starts to glow in the invis-ible infrared end of the spectrum. Increase the heat and its glow moves toward blue, crossing the visible spectrum. It was known for almost a century that the amount of energy released is re-lated to the wavelength of the radiation.

If we could magnify some material solid a quadrillion times we might see it as a large array of atoms or molecules, each ex-erting forces on its neighbors, each pulling or being pulled. We might think that each atom or molecule is in equilibrium, but as heat (energy) is applied each will vibrate about its equilib-rium position. The temperature of the solid is a measure of the average kinetic energy of the vibrating molecules—the higher the temperature, the larger the vibration of molecules.

In the early months of 1900, soon after the electron was discovered, Max Planck wondered why radiation changes color continuously from red to blue as temperature rises, and hypothesized that energy does not exist as a constant electromagnetic wave, but rather as individual quantifiable units, just as matter does. When Planck tried to fit the classical theory of radiation (in which the oscillators could support a continuous array of energy values) to results such as the shift from red to blue, he found that atomic oscillators could not achieve continuous energy levels, because they were required to oscillate by discrete amounts of energy. His idea was to partition radiation and energy into a large number of discrete parts to deal with them as proportional to the frequency of radiation manifesting as color. Energy was given by $E = vhn$, where v is the frequency of the oscillation, h is Planck's constant (6.625×10^{-34} joule-seconds), and n is an integer called the *quantum number.* Hence, the amount of energy radiating from the black body only comes in n very small bundles of equal amounts of energy, namely hv, a quantum of energy, later to be known as a *photon.*

Planck's constant h is extremely small, so small that the photon fools us into thinking that light is continuous when in fact it is as discontinuous as water. Not only is matter discontinuous, but so is light, the means to see matter. We might see an arrow silently moving from its bow to its target, imagine that its path is continuous, and even think that it can smoothly move with time across its path. But even with the sharpest focus through our most precise instruments, we can never see that infinitesimal discontinuity and must rely on the impressions we get from those deceptive quanta that pull together and make us believe that space and the arrow's flight through it are both continuous.

At the time of Planck's most brilliant discovery he was forty-two and a professor of theoretical physics at the University of Berlin, studying thermodynamics and the distribution of energy by wavelength. He had a kind face with low ears and receding hair at the temples, but he was still a rather handsome man. Though he confidently talked about his "elementary quantum of action" at the December 14, 1900, meeting of the Berlin Academy of Sciences, he had his own doubts about going so far against classical physics. At his Nobel Prize acceptance speech he said, ". . . the quantum of action must play a fundamental role in physics, and here was something completely new, never heard of before, which seemed to require us to basically revise all our physical thinking, built as this was, from the time of the establishment of the infinitesimal calculus by Leibniz and Newton, on accepting the continuity of all causative connections. Experiment decided it was the second alternative."

FROM ZENO'S TIME to the present much of physics has been about reducing the world down to fundamentally discrete, countable particles; yet it is difficult to imagine fluid motion, time, space, and the propagation of light as discontinuous. Louis de Broglie wrote, ". . . physicists have gradually reached the conviction that the continuous character of solids and fluids is illusory, and that in reality they consist of atoms in motion, while it is only the obtuseness of our senses which prevents us from perceiving this ultimately corpuscular structure of Matter, and causes us to suppose it continuous instead." De Broglie was onto something strange when he was still a doctoral student at the Sorbonne in 1924. Physicists knew that electromagnetic waves could be described as particles. In 1887 Heinrich Hertz, the man well known for detecting Maxwell's invisible electromagnetic waves—waves that reflect, refract, diffract, and travel at the speed of light—discovered that an electric spark brightens when bombarded with ultraviolet light, a phenomenon called the *photoelectric effect.* Later—after the discovery of the electron in 1897—more sophisticated experiments performed by the Austro-Hungarian physicist Philipp Lenard determined that the frequency of the light beamed on the metal controlled the energy of the electrons emitted by the metal; that there was a threshold frequency below which no electrons are emitted; and that the intensity of the light didn't matter if the frequency was below the threshold. He showed that the energy of any electron emitted from light depended only on its frequency, i.e., its color.

Einstein had an explanation for the photoelectric effect. In a paper he completed on March 17, 1905, he suggested that the energy of light is distributed in space discontinuously, and that when a "particle of light" of high enough frequency penetrates a metal, it hits an electron, transferring its energy to that electron, which in turn uses that energy to escape. The higher-frequency photons have more energy, but higher intensity just means more photons are emitted. If the frequency (energy) of the light (photon) is below the threshold, it will not be able to free an electron from the metal, and since all the photons of that particular frequency have the same energy, increasing the intensity (the number of photons below threshold) will not make a difference. Einstein wrote, ". . . in the propagation of a light ray emitted from a point source, the energy is not distributed continuously over ever-increasing volumes of space, but consists of a finite number of energy quanta localized at points of space that move without dividing, and can be absorbed or generated only as complete units." According to the Cambridge astrophysicist John Gribbin, "That sentence marks the true beginning of the quantum revolution." Perhaps he's right; in any case, Einstein was belatedly awarded a Nobel Prize for this 1905 idea in 1922.

In 1924 de Broglie began to think that electrons could be described as waves. Putting together two simple equations that had already been known, he derived a very simple relationship between p the momentum of light particles, λ the wavelength of electromagnetic radiation, and Planck's constant h. His revelation was not simply that $p\lambda = h$, but that it applied to electrons as well as photons. This opened a new way of thinking about not only the electron but also about the atom itself, for if the electron has a wavelike orbit around the nucleus, then the energy levels of the electrons must match the wave's harmonics. This means that the number of waves of the orbit must be an integer. The orbit can't be one-and-one-half waves, because the wave must return to itself.

De Broglie assumed that nature loves symmetry and, extending Hertz's discovery, suggested that, on the atomic and subatomic levels, energy and matter both behave as though they are particles and waves. However, there is a significant difference between a particle and a wave in the way energy is stored. The energy of a particle is concentrated in its mass; the energy in a wave is spread throughout the wave. The only way out of the ancient rivalry between particle (Newton) and wave (Young) characterization of light was to assume that it was both: it has properties of particles and waves, without being strictly considered one or the other. A particle is accompanied by a wave, and every wave is accompanied by a particle in motion. Light, which had always been thought to be continuous, was now thought to be composed of discrete *equal* quantities, or quanta proportional to the frequency of the radiation (equal to $h\nu$).

· 14 ·

There's No Next, but What's Next?

Imagine Superman being able to see vast distances and also able to magnify tiny, almost infinitesimal specks with superpowers that enable him to elude the uncertainty principle and look at the innermost secrets of space without distortion. He looks at the outer regions of our galaxy and senses the curves in space near large gravitational fields. But when he looks deeper and deeper into a speck of sawdust, he begins to see space below the quantum levels and finds it to be in a turbulent sea of foaming randomness, an image in sharp contrast to the smooth dimples in the nylon of the general relativity trampoline.

Unlike general relativity's compatibility with Newtonian mechanics, there is an abrupt difference between general relativity and quantum mechanics at Superman's powers of magnification, where the general relativity model of space breaks apart, suggesting a search for a more general natural mathematical model, one that would include both general relativity and quantum mechanics. In the geometry of general relativity, the universe is bumpy with dimples wherever there is a mass, but those bumps are smooth.

In the microscopic fabric of space we do not see smoothness; rather, we don't see anything but uncertain quantum fluctuations of gravitational fields so violent that they average out to zero gravity. And the Heisenberg uncertainty principle tells us that the closer we look, the more violent the fluctuations of the gravitational field become. But in our general relativity geometry, gravity corresponds to dimples or the warping of space, so the micro-microcosmic fabric of space must be fiercely turbulent, undulating in frenzied disturbance, contradicting the general relativity notion that space is fairly smooth.

In the wake of general relativity and quantum mechanics the indivisible-point model of the universe could no longer continue to represent our universe. An elegant theory developed in the early 1970s, the *standard model* of particle physics, which describes the strong, weak, and electromagnetic fundamental forces and assumes that all matter is made from fundamental particles. It is consistent with both quantum mechanics and special relativity (which does not take the gravitational force into account) and has been verified by experimental tests. But the standard model does not describe gravitational interaction, the oldest known force and the weakest of the four forces of nature.

In the mid-1980s physicists turned to a new theory, a new quantum theory consistent with gravitation, with uncontrolled enthusiasm. Brian Greene said that when he was a graduate student at Oxford in 1984, first-year graduate students had an "electrifying sense of being on the inside of a profound moment in the history of physics. . . . A number of us consistently worked deep into the night to try to master the vast areas of theoretical physics and abstract mathematics that are required to understand string theory."

String theory, it was called. Its first of several phases came in 1968, when the Italian physicist Gabriele Veneziano learned that a formula discovered by the eighteenth-century mathematician Leonhard Euler perfectly matched data described by the strong forces of interacting particles. Any formula perfectly matching collected data begs for a reason. That reason soon came from Chicago, Stanford, and Copenhagen, when Yoichiro Nambu, a Japanese-American, Leonard Susskind, an American, and Holger Bech Nielsen, a Dane, discovered that Euler's function described elementary particles as very tiny vibrating strings.

The theory of general relativity describes the fundamental force of gravitation applied to large-scale structures such as stars and galaxies, whereas quantum mechanics describes the remaining three fundamental forces—the electromagnetic, strong nuclear, and weak nuclear forces—that function on a microscopic scale. Moreover, there is that annoying problem of conforming the smooth texture of general relativity's space-time with the violent behavior of the universe at high magnification. The object of the game is not to worry about whether or not the texture of space is smooth, but rather to build a single new theory that retains the many substantive ideas of both the general relativity and quantum theories and to compatibly explain the nature of all four fundamental forces. The new theory should include a description of a spatial fabric that is smooth at the macroscopic level and violently unsmooth at the micro-microscopic level, so that general relativity could accommodate quantum physics. String theorists optimistically speculate that they may someday have such a theory.

Now these strings are indeed tiny, so tiny that even magnifying one 10^{20} times would bring it to about the size of the nucleus of a hydrogen atom—unimaginably small. But when the standard model of particle physics is replaced by a theory that includes those tiny vibrating loops of strings, a promising theory emerges, a theory that encompasses general relativity and quantum mechanics, one that promises to explain all four fundamental forces of particle physics.

What are these vibrating strings? As tiny as they are they must be made of something. But they cannot be made of atoms, for they are already deep within the matter of atoms. Though they cannot be seen, even with the most powerful instruments, at large theoretical magnifications we imagine them as breadthless points. But they are not breadthless; if they could be magnified to the size of a pea, and if we could slow time down so that a nanosecond would take a year, we would see them as rapidly waving loops of something that looks like string. The advantage of the string disguised as a point is that it encompasses the idea of elementary particle and fundamental forces in one package.

Take the *spin* of an electron. Quantum mechanics assumes that the electron sort of both rotates and revolves along a path, though its position along that path can only be probabilistically known. Not only does it spin at a fixed rate, but that rate is also the same for all electrons in any atom. Other subatomic particles in the same family, such as quarks, also spin at that same rate. The spin creates magnetic properties and is one characteristic that is mimicked by strings—spin is characterized by a string's vibration pattern.

Special relativity tells us that energy and mass are interchangeable. And since gravity is determined by mass, we find that a string's activity gives it its gravitational force. In fact, the other three fundamental forces of nature (the electromagnetic, weak nuclear, and strong nuclear) may also be connected to the particular pattern of vibration.

So here they are, strings like strings on a guitar that can vibrate in a huge number of different (yet similar) wave patterns, mainly characterized by the integral number of waves—the higher the number, the more active the vibration and hence the higher the kinetic energy. Each different particle is distinguished by the string's vibration pattern.

It's possible that these strings are truly fundamental in the sense that they have no continuant parts, and that they are the true indivisibles from which all matter is made. Perhaps they are what Brian Greene called, "the last of the Russian *matrioshka* dolls," in which case the question of what they are made of makes no sense: they can't be made from anything but themselves, for otherwise they would not be the last *matrioshkas*. Maybe strings are the last word on the odyssey of the motion paradox. Or perhaps they're not.

By 1995 there were five distinct versions of string theory that seemed to be connected and that seemed to be special cases of the correct theory of everything, a theory that purports to explain all phenomena behind all forces of nature and give a picture of the fundamental "things" that make up the universe.

One version predicts that the number of dimensions that the

universe possesses is not four, but twenty-six. Other versions have the number of dimensions down to ten. We observe four, so where are the other six dimensions? One way of thinking about those extra dimensions is that they are *fibers* on the four-dimensional space; each point of the four-dimensional space has a fiber and each fiber is a six-dimensional mathematical entity. These fibers are so tiny and curled that we imagine them as one-dimensional lines rising from the points of our four-dimensional space. But the lines really represent six more dimensions, in order to incorporate enough information about how the space is to react to the fundamental forces and how events passing through that space are to behave. In the string-theory model, space is made not from points (mathematically characterized as a list of address numbers), as it is in the standard model, but from fibers mathematically manifested as matrices.

Why the extra dimensions? Think of how a guitar string oscillates in two dimensions. The pressure of a finger on a fret changes its frequency, but so does the wood of the guitar that sits in one higher dimension and resonates in sympathy with the plucked string. The one-dimensional guitar string transmits its wave effect to the two-dimensional wood surface sitting in our normal three-dimensional space. As the guitar resonates with its string, so do the four dimensions of space-time resonate with the tiny curled-up six-dimensional micro-microscopic fibers whose vibration patterns manifest as masses and forces. We cannot visually see those half-dozen compact dimensions, and even in physics they are so small that they are undetectable, but we can feel them as the masses and forces of elementary particles and experience them as gravitational and electromagnetic forces. The original concept of *dimension* still holds, even if it means twice measured—once by sight and again by a totally different sense.

In 1984 the American physicists Gary Horowitz, Andrew Strominger, and Edward Witten picked up a mathematical object that American mathematician Eugenio Calabi had originally conjectured in 1957. The object had a six-dimensional abstract space with just the right kind of metric (the yardstick by which to measure) and the very symmetries needed to model the symmetries of fundamental particles.

Calabi's space came from differential and algebraic geometries and a specific classification of algebraic surfaces coming from the roots of polynomial equations. He conjectured that a particular class of six-dimensional spaces has remarkable symmetries and a so-called *Ricci-flat metric* (a particular kind of yardstick by which to measure within the space). In 1977 the Chinese-American mathematician Shing-Tung Yau had proved Calabi's conjecture.

The class of surfaces, so-called "Calabi-Yau manifolds," had the right properties—the right dimension, the right metric, the right topology, and just the right symmetries—to fit properly into a theory that would closely model the behavior of elementary particles viewed as string vibrations moving through it. Our imagination of space is built from our experience with space, but we have no experience with space in the tiny neighborhoods of strings the size of 10^{-35} of a meter. However, one way to visualize something in a dimension higher than three is to project its image onto a lower dimension; for example, a photograph of an apple is a projection of a three-dimensional object onto flat photo paper. Another way is to slice it by a series of lower-dimensional spaces. For example, if the apple were thinly sliced, the sequence of slices would give us the impression that whatever it is was small and circular in the first slice, circular and growing larger in radius for the next few slices, circular and becoming smaller in radius for some of the following slices, and small and circular for the final slice. Reason then tells us that the apple must have been spherical in three dimensions. Such slicings are performed mathematically by holding variables fixed.

We come to the Calabi-Yau manifold in the same way. The illustration below is a two-dimensional projection of a four-dimensional cross section of a six-dimensional Calabi-Yau manifold. That's the only way we can *see* it on a page of a book.

Andrew J. Hanson, Indiana University

Recall that the worldline is the graph of an event as time passes. If we draw the worldline of a one-dimensional string in space-time, it will sweep out a two-dimensional surface in space-time known as a *worldsheet*. For example, we must imagine that the vibrating pattern of the string will induce waves on this worldsheet—possibly electromagnetic waves or possibly gravitational waves. Being a loop, the string will sweep out an undulating pipelike-figure worldsheet. But the real question is what does space look like up close, very up close? We once thought of it as a continuum of points and that every so often there would be a molecule with its atoms, an atom with its nucleus, electrons, and protons, etc. But what about all that empty space between those molecules of matter, between those atoms, between a nucleus and its electrons, or between quarks? What does empty space mean? What does the word *between* mean in the context of points of space? The string theory view is to look at it as though the first four dimensions were space-time, but that every point in space-time is really some Calabi-Yau manifold.

Those vibrating strings move through the Calabi-Yau manifold, up and down the six dimensions. Motion in such a universe is far more complex than anyone previously thought. Zeno never imagined that his arrow would have to make its trip not only in the continuous universe of space, or space-time, but also up and down the six dimensions of whatever Calabi-Yau manifold sits at each point that the arrow passes. He never thought Achilles had to catch up with the tortoise by such a very long journey. Of course, the extra dimensions are so small it takes no time for either the arrow or Achilles to move through them.

If string theory is truly a model of the universe (remember that the model is still speculative), then Zeno's paradoxes of continuity are far deeper than he and we had ever expected. Everything we do, every movement we make is really an illusion—the illusion that we move only through three dimensions.

Zeno and Parmenides had suggested long ago that our conceptions of reality are fantasies, illusions. Einstein put a twist on that idea, saying that truth and reality are justified if the concepts of reality can be correlated with experience. "We are free," he wrote, "to choose which elements we wish to apply in the construction of physical reality. The justification of our choice lies exclusively in our success." If our models work, we are justified in using them to understand reality. But we should never confuse the model with physical reality.

Zeno would heartily agree that the mathematics applied to his paradoxes makes sense and that it can pinpoint with absolute accuracy when and where any one of his phenomena will happen. But still, the more mathematics proves and the more physics shows about the world, the more paradoxical its motions seem.

The One Stream

 The causes and nature of human consciousness had been persistent topics of debate since Plato's time, energetically revived when Descartes questioned his own existence. At the end of the nineteenth century, the psychologist William James applied continuity of time to his investigations of what he called the *stream of consciousness,* arguing that it is impossible to stop any thought for introspection before it reaches a conclusion. If, with some luck, the thinker is "nimble enough to catch it, it ceases forthwith to be itself." It seems that a conscious thought evaporates before it can be examined, like "a snowflake crystal caught in a warm hand." Any attempt to freeze the continuous stream of a human's conscious thoughts is as pointless as stopping "a spinning top to catch its motion, or trying to turn up the gas quickly enough to see how the darkness looks." These are "as unfair as Zeno's treatment of the advocates of motion, when asking them to point out in what place an arrow is when it moves."

We "may live through a real outward time, a time known by the psychologist who studies us, and yet not *feel* the time, or infer it from any inward sign." Perhaps consciousness, itself, is discontinuous, "incessantly interrupted and recommencing (from the psychologist's point of view)?" Surely it is interrupted by sleep and dreams; yet it *seems* so continuous. Is it an "illusion analogous to that of the zoetrope? Or is it at most times as continuous outwardly as it inwardly seems?" William James had no answers to these questions and claimed there were none.

We may not be able to tell whether or not conscious thought is continuous, but we do know that the complex bundles of signals perpetually collected from all human senses are tidily synchronized and recorded to form what we call *consciousness*.

Take sight. The zoetrope, a nineteenth-century parlor-room toy that gives the illusion of motion, is nothing more than a spinning cylindrical drum containing slits and no more than a dozen still images of a person. Each image is very much like the next, except for a slight difference in anatomical position. When the succession of discrete images is viewed through the slits of the turning drum, the viewer sees the images fused into a dynamically moving picture—a person in motion.

Still images in rapid succession are interpreted as real continuous motion. How does this happen? The celebrated nineteenth-century physicist Hermann von Helmholtz, in one of the great contributions to medicine, his *Handbook of Physiological Optics*, thought that the eye held one image just long enough for the next to take over. Something of the sort actually does happen in the retina; look at a black spot on a white background for a few seconds and then turn away. The black spot will linger for a few seconds more. This is even more pronounced when we see a spot of bright light in a dark room long after the spot of light is extinguished. The spot is temporarily *burned* onto the photosensitive retina. But we now know that the coordination of discrete visual images—real sight—takes place in the visual cortex, not in the eyes, so the question remains: How is it that a rapid succession of still images is construed as a moving picture seamlessly flowing in time?

Is there some biological necessity to see continuous movement? A frog is uniquely capable of catching bugs because it sees only the movement of its tiny prey, undistracted by inconsequential surroundings. Humans do not need to catch flying insects with their tongues, but need far more than just visual receptors to sense movement. Humans were once capable of hunting and fishing as well as watching out for ambushing saber-toothed tigers or stealthy reptiles. But why the need for continuous movement? Wouldn't staccato motion be enough to protect us from wild beasts and enable us to farm and live as we always have?

Perhaps the real world that we live in is truly disjointed, where every movement behaves as though it is a flickering Max Sennett Keystone Comedy. If it were, would we know it? Or would we simply wrap what we see into a blanket of personal sensations relative to what we expect to see? If the brain is truly the thing doing the seeing, then it hardly matters what we see as long as we give enough information to the brain to interpret reality. Helmholtz experimented with prism glasses, which turned his field of vision upside down. It didn't take long for his brain to compensate, with the help of his physical existence in the real world, and put the world right side up. So shouldn't our magnificent brains be capable of adjusting the flickering world of Keystone Comedy so it appears smoothly continuous, as long as we give it the time to do so?

And what about the other way around? What if the real world behaved smoothly and we could only see it through stroboscopic light? Would we feel discomfort at seeing a disjointed succession of images? Would we know the difference, or would we simply adjust as we did to foods that had bothered us in infancy?

OUR PERCEPTION OF Zeno's continually moving arrow may have nothing to do with its true movement. And yet, the mathematics we installed to model his arrow's flight must also model our sense of that flight. So the paradox stands between the crossroads of reality and our perception of it. The same holds for the dichotomy or the Achilles and tortoise paradox, in which Zeno claims that an infinite number of events must take place before anything is accomplished.

The mathematical biologist D'Arcy Thompson, who gave us the splendid idea that biological growth and form can be described through mathematical relations, claimed that "the harmony of the world is made manifest in Form and Number, and the heart and soul and all the poetry of Natural Philosophy are embodied in the concept of mathematical beauty." The English physicist Sir James Jeans once wrote: "From the intrinsic evidence of his creation, the Great Architect of the Universe now begins to appear as a pure mathematician." And Galileo described the universe this way: "This grand book is written in the language of mathematics, and its characters are triangles, circles, and other geometrical figures."

These celebrated quips seem to lose some of their punch after the German number theorist Leopold Kronecker's one-liner, "God created the integers, the rest is the work of man." Did he mean that mathematics strayed from the integers when man introduced infinity to the mind? Was this drift from God's creation responsible for mathematics' wandering from physical reality?

David Hilbert said, "And the verdict is that nowhere in reality does there exist a homogeneous continuum in which unlimited divisibility is possible, in which the infinitely small can be realized. The infinite divisibility of a continuum is an operation which exists in thought only, is just an idea, an idea which is refuted by our observations of nature, as well as by physical and chemical experiments."

In these lines, Hilbert is suggesting a friendly quarrel between the concept of infinity—the principal ingredient of continuity—and the perceptual world as mathematics brushes close for a clear look. And a friendly quarrel it is, because like very young siblings, they still need and nurture each other. We know the physical world by how we perceive it, and we perceive it by how we measure it. The moment we attempt to measure, by ruler, scale, gauge, compass, or thermometer, we are sanctifying number. In that moment we are assuming continuity through a precise correspondence between what we see and what is really there to be seen. But we don't measure with infinitesimal instruments and so must be content with relatively rough estimates of reality.

For each paradox in Zeno's quiver, we have an answer—*continuity is merely a conscious impression, a fabrication of the mind elevating illusion to reality.* Though mathematicians may try to explain the paradoxes by logical models of motion phenomena such as algebra or infinite series, they miss the target: to give a phenomenological explanation of the unavoidable sense of harmony between the fantasy of time and the continuously flowing universe. Yes, they can tell us precisely where the arrow is, when Achilles will overtake the tortoise, or when we will come to the other side of a room, but they cannot tell us *why* without bending our perception of space to fit our inflexible intuition of time's continuous nature.

Ask why Achilles overtakes the tortoise and the response will inevitably be, "Because we see that the algebra tells us that it happens when. . . ." Ask again, and the response will point to a mathematical model. We know that the model—the one constructed to give the answer—is based on the continuous nature of the real-number line, which cannot precisely imitate a phenomenological nature of real matter composed of atoms with their excited electrons permitted to change orbit only by discrete jumps and their energies changing by discontinuous quanta packets.

Ask why the earth orbits the sun according to Kepler's laws and the answer will be *universal gravitation.* Hurrah, universal gravitation tells us that two bodies attract with a force inversely proportional to the square of the distance between them. But *why?* The answer is clear. Force is proportional to acceleration, that's the old $F = ma$ formula relating force to mass and acceleration. But these are merely physics terms suggesting—from experiment, of course—that there is a pull that can be felt and measured and an increase in speed that can also be measured. All we need to know is that the pull increases as the acceleration does. Then $F = ma$. But *why?*

Why does the ball continue on its trajectory when it leaves the hand that throws it? The answer invokes a combination of laws—an object will continue to move in uniform motion unless acted upon and $F = ma$. Whenever we ask why, we are cornered into a mathematical formula that was initially constructed by relating physical phenomena.

Why do electric and magnetic fields have such a strong symmetry, in the sense that a changing electric field induces a changing magnetic field, and vice versa? Why are they both just two forms of radiation? Why is mass just another form of energy? The list goes on. Ask, and you will be given some mathematical model as the answer.

Mathematics has done a pretty good job of formulating the laws of physics; does that give it the right to be at the end of the line of *whys*? Why does Achilles overtake the tortoise? Because a geometric series with ratio less than 1 converges. Satisfied? No. We are more satisfied after watching Achilles beat the tortoise. Why is mass just another form of energy? "Because $E = mc^2$," says the physicist. Satisfied? No. The horrific bombings of Hiroshima and Nagasaki were more convincing.

IN 1960, THE Nobel Prize–winning physicist Eugene Wigner wrote a classic essay entitled, "The Unreasonable Effectiveness of Mathematics in the Natural Sciences," in which he told the story of a statistician showing a paper on population trends to a friend who knew little mathematics. The friend pointed to the symbol π and said, "Surely the population has nothing to do with the circumference of the circle."

Wigner's story embodies a few points: first, that problems

involving the real world translate into weirdly unexpected mathematical notions; second, that cognition may be the source of physical concepts; and third, that "the enormous usefulness of mathematics in the natural sciences is something bordering on the mysterious and that there is no rational explanation for it." How else can we explain why mathematics describes so many of the physicist's raw encounters with nature? "We do not know why our theories work so well," Wigner wrote. "Hence, their accuracy may not prove their truth and consistency."

As an example Wigner used the law of universal gravitation: "The law of gravity which Newton reluctantly established and which he could verify with an accuracy of about 4 percent has proved to be accurate to less than ten thousandth of a percent and became so closely associated with the idea of absolute accuracy that only recently did physicists become again bold enough to inquire into the limitations of its accuracy." It seems that Newton must have stumbled onto his law from crude measurements and empirical hunches to express a remarkably simple mathematical formulation that turned out to give an amazingly accurate description of motion—not the cause—and its influence on surrounding masses. That stumbling, according to Wigner, was a miracle. And because of it, men have walked on the moon; robots have played in the red sands of Mars and probed comets and asteroids; and cameras that have been flown to the edge of the visible universe have sent pictures back to Earth.

It would be wonderful to have an answer that would explain away the paradox, an argument about continuity perhaps, or a trick to untangle the infinitesimal fabric of the continuous line. But our only answer seems to still be Zeno's. He said it twenty-four centuries ago. If we were to ask him why we see the arrow leave the bow and hit its target, he would still respond, "Mere appearance of change. Motion is an illusion," and possibly add, "Now that you've had more than twenty-four centuries to ponder the problem, you know that even matter is nothing more than energy, and vice versa. Nothing has changed. The external world may be material known only by our senses giving the illusion of color, smell, feeling, and motion."

Acknowledgments

When Bertrand Russell said, "We want to have ten fingers and two eyes and one nose," he was referring to the goal that logicians had in mind when they first set themselves the task of defining number, hoping that the definition would apply in the right way to common objects. I had many fingers, eyes, and noses while writing this book, thanks to the help of many friends and colleagues who encouraged me in conversations, diligently read rough versions of the manuscript, and made constructive suggestions and crucial corrections. First, I thank my wife, Jennifer, who read the complete manuscript after listening to me read many chapters aloud. Without her practical suggestions this book would not be in print. She is my greatest support, anchor, and inspiration.

I am exceedingly grateful to Stephen Morrow, executive editor at Dutton, for brilliantly transforming my ambiguous pitch of a rough idea into a tidy, meaningful, extraordinary plan. He did a colossal job in reorganizing the flow of the last half of the book. Also to Jeffrey Galas, my editor at Dutton, for his insightful suggestions and careful editing, which always seems to miraculously transform what I say into what I mean to say. Without his astute editing and clever suggestions, this book would be unreadable.

Very special thanks go to all those who read large parts of the manuscript and contributed constructive suggestions and corrections to make it make sense: John MacArthur and Travis Norsen of Marlboro College made extensive corrections to drafts of the text on relativity and quantum theory, while illuminating many examples. James Callahan of Smith College generously spent many hours working with me on detailed corrections, giving significant advice for the final draft. Emily Grosholz of Pennsylvania State University, Mark Huibregtse of Skidmore College, and Robert Perlis of Louisiana State University also gave considerable time, contributing expert advice and corrections. Sorina Eftim of Johns Hopkins, Timothy David Hirrel, Peter Merideth, and Mark Ollis of Marlboro College also made significant corrections. My very good friend Jay Birjepatil of Marlboro College read portions of a late draft and gave friendly advice, but his most significant contribution was—as always—his luminous intellectual presence. I am forever indebted to all these special friends and brilliant advisers.

I am especially thankful to Ray Bates, The British Clockmaker of Newfane, Vermont, for kindly inviting me into his workshop and teaching me so much about the mechanical workings of ancient clocks; to Willene Clark of Marlboro College for her neighborly support and her expert information about medieval glass; to Udo Schubach for his friendship and help in procuring documents at the Munich Museum; and to

Edward Adelson at the MIT Department of Brain and Cognitive Science, for bringing me up to date with information on how the brain processes sight signals of moving objects. My remarkably perceptive ninety-one-year-old mother-in-law, Anne Joffe—a person I love dearly—listened to me read parts of the original manuscript aloud and asked perceptive questions that helped shape the final draft.

The wonderful listeners and critics in Sylvie Weil's writer's soiree, Arlene Distler, Anne and Tony Gengarelly, Michael Kennedy, Jennifer Mazur, Franklin Reeve, Laura Stevenson, Sylvie Weil, Eric Weitzner, and Anne Wheelock, have given me the perfect forum for experimenting with writing styles and voices.

To my supportive friends who have not read the original manuscript but who have listened to my many stories and are constant inspirations for executing good work: Tadatoshi Akiba, William Bown, and Ian Stewart. To Evan Johnson, my young neighbor and tutee, for helping me learn how to teach mathematics and science to a teenager. As always, I am beholden to the enormously competent Marlboro College library staff, Mary White, Elsa Anderson, Radmila Ballada, and Elizabeth Dolinger, for gracious and expert assistance in finding cross-referenced material and interlibrary loans with such good cheer, often after hours. To Jennifer Bryan and the staff of Special Collections & Archives Division of the Nimitz Library at the Naval Academy for their kind hospitality during my research into the life and work of Albert Michelson.

I would know next to nothing without my students; they teach me so much. Students in my "Motion and Number" course at Marlboro College shaped my thoughts of how to present the material in this book. Those students are: Amber Nuite, Ambrose Sterr, Eliot Gluckman, Jessie McNabb, Joelle Montagnino, Joshua Lande, Samantha Williams, Emily Cahill, Evan Mehler, and Brian Reed.

Thanks to Bruce Cole's generosity and his cherry tree, this project was mostly accomplished at an exquisitely beautiful and organized desk, setting the relaxed mood and spirit so necessary.

To Barry and Gretchen Mazur for their constant support. To Catherine Mazur Jefferies, Tom Jefferies, Tamina Clark, and Steven Clark. And, too, I am grateful to my granddaughters, Sophia and Yelena Mazur Jefferies, and Lena Clark for showing me how to learn new things.

Last, I am indebted to the many nineteenth- and twentieth-century historians and scholars who translated and edited original texts, works, and letters into modern language. I relied on them for the truth of history.

Notes

11   *to attend the great festival*
      We know very little about the life of Zeno. Aristotle credited him
      as the inventor of dialectic, the type of argument that involves
      claim and contradiction to isolate truth, because his tactic was to
      draw objectionable conclusions from his opponents' theories. His
      visit to Athens and a small part of his philosophy is recounted by
      Antiphon in Plato's dialogue *Parmenides*. We get a bit more biog-
      raphy from Diogenes Laertius's *Lives of Eminent Philosophers*,
      written more than 700 years after Zeno's death. Still, if we assume
      that Diogenes had access to a continuous collection of resources for
      biographical research—lost to us—we may accept his picture as
      reasonably accurate.

12   *He called out to Zeno*
      Zeno of Elea is often confused with the more famous philosopher
      Zeno the Stoic, who lived between 340 and 265 BCE.

12   *"Yes," replied Zeno*
      This dialogue is reasonably close to what Antiphon claims to have
      taken place. See Plato's *Parmenides*, *The Collected Dialogues of
      Plato*, 127 d-e.

12   *being, continuity, and motion*
      Zeno's argument may be confusing, but to get a clear idea I recom-
      mend reading the full argument in Plato's *Parmenides*, starting on
      line 137 and ending on line 142.

13   *neither ceases to be nor comes to be*
      Ibid., 163d.

13   *never catch them moving*
      From Oliver Sacks, "Speed: Aberrations of Time and Movement,"
      The New Yorker, August 23, 2004.

14   *how to raise virtuous citizens*
      Diogenes Laertius.

14   *politics, the arts, and philosophy*
      At the height of its power, the population of Athens did not exceed
      300,000. More than half of this number were slaves and foreigners.
      About 200,000 were women and children.

14   *an experiment in democracy*
      By *people* we mean *free people*.

16   *was the construction of the regular pentagram*
      A *regular* pentagon is a polygon with five equal sides.

16   *all the angles of construction of the regular pentagram*
      T.L. Heath, *Euclid*, Vol. IV. New York: Dover Publications, 1956, p. 10.

16   *along with its powerful numeric and geometric qualities*
To replicate the pentagon connect each of the five corners to each of the four remaining corners by straight lines. A smaller pentagon will appear inside the original pentagon. To make larger pentagons extend each of the five sides by straight lines until they meet the extensions of each other. This will give a pentagram. Connect the five corners of the pentagram by straight lines. A larger pentagon will appear. In this way a single pentagon can be replicated infinitely by smaller and smaller pentagons and also by larger and larger pentagons. The universe could be filled with pentagons generated by the original pentagon.

17   *not to be coined until the nineteenth century*
It is first found in *Archive der Math. und Physik*, Vol. IV, 1844, 15-22.

18   *solution to the equation 4x + 20 = 4*
T.L. Heath, *Diophantus*, 2d. Ed. Cambridge, 1910, p. 52.

18   *to give the square root of 2*
Numbers that cannot be written as a ratio of two integers are called "irrational," that is "not rational."

18   *in its endeavor to understand the world*
Bertrand Russell, *Scientific Method in Philosophy*. London: Open Court, 1914, p. 164.

20   *Empedocles says it this way:*
This is a translation of an Empedocles poem (Diels-Kranz B-text listing B23) from *Early Greek Philosophy*, translated and edited by Jonathan Barnes. London: Penguin, 1987, pp. 167-8.

22   *with her tears the mortal fountains*
This is a translation of Sextus Empiricus (Against the Mathematicians X 315—Diels-Kranz B-text listing B6). Ibid., pp. 173-4.

22–23 *More directly, he says*
Translation from Clement, *Miscellanies*, V. viii. 48.3 (Diels-Kranz B-text listing B38). Ibid.

25   *without limit before it gets there*
Ibid., p. 181.

25   *all this in a book*
In the fifth century BCE, books were read aloud by the author, not sold to the public as they are today.

25   *the assumption of plurality and motion*
Dictionary of Scientific Biography. New York: Charles Scribner's Sons, 2000.

25   *a central hearth over a stone floor*

This was a special room called the *andron* that was reserved for men only, although female servants, musicians, and dancers who performed for symposia were permitted to enter.

27   *no such thing as a next number*

Take the number $\frac{1}{2}$ as an example: What is the next larger number beyond $\frac{1}{2}$? Is it $\frac{3}{4}, \frac{5}{8}, \frac{9}{16}$ or $\frac{17}{32}$, or some other number of the form $\frac{2^n+1}{2^{n+1}}$? Every number of the form $\frac{2^n+1}{2^{n+1}}$ will be larger than $\frac{1}{2}$.

It does not matter what n is. Moreover the larger n is, the closer $\frac{2^n+1}{2^{n+1}}$ is to $\frac{1}{2}$, and there is no n that would make $\frac{2^n+1}{2^{n+1}}$ the next larger number than $\frac{1}{2}$.

27   *the substitute to the genuine article*
Tobias Dantzig, *Number: The Language of Science*. New York: Pi Press, 2005, p. 132.

28   *hitherto accessible to our observation*
The quote is a translation of a passage in David Hilbert and Paul Bernays, *The Foundations of Mathematics*, from the German found in Stephen Cole Kleene, *Introduction to Mathematics*. Princeton: Van Nostrand, 1962, pp. 54-5.

32   *ranging from love to medicine*
For a catalog of his works see the above endnote, V 21-28, p. 465-475. We should note that a book in Aristotle's time was more like a chapter of a modern book. According to Diogenes Laertius, he wrote 445,270 lines, which, in today's print would be roughly equivalent to between twenty and thirty books, depending on how many words he fit on a line. The awe comes in reviewing the breadth of topics of his catalog.

33   *a liquid or a hot thing becoming cold*
Aristotle, *The Physics*, p. xv. (Quotes on the next four pages are from this source.)

39   *will be finished in just one hour*

That's because the sum of the sequential powers of 1/2 equals 1.

40   *independent of distance measurement*
In other words, properties such as stretching without tearing or gluing.

42   *which assumption is false*
Aristotle, *The Physics,* VI. ix. 240a 5.

42   *being kept in motion by something*
Aristotle, *The Physics,* VII. i, 241b 24.

46   *only by the death of the guilty*
Edward Gibbon, *The Decline and Fall of the Roman Empire*, Vol. II. New York: Washington Square Press, 1962, p. 531.

46   *help to them when they were sick*
From the Prologue of Chaucer's *Canterbury Tales*, translated by Nevill Coghill. Baltimore: Penguin, 1958, p. 17.

47   *a wheat field in Clermont, France*
Written records of Urban II's speech date from several years after the speech was delivered, so there is no way to verify what the pope actually said. However, there are several versions of the speech that differ in language and yet agree in intent. Dana C. Munro compares the various texts in "The Speech of Pope Urban II at Clermont, 1095," *The American Historical Review*, vol. II (1906): pp. 231-42.

47   *in the kingdom of heaven*
Ibid. This may also be found in Daniel J. Boorstin, *The Discoverers*. New York: Random House, 1983, p. 118.

48   *under penalty of excommunication*
A Source Book in Medieval Science, edited by Edward Grant. Cambridge, Massachusetts, Harvard University Press, 1974, p. 42.

49   *in founding modern astronomy and physics*
St. Thomas Aquinas, *Commentary on Aristotle's Physics*, translated by Richard J. Blackwell, Richard J. Spath, and W. Edmund Thirlkel. London: Routledge & Kegan Paul, 1963, p. xviii.

50   *teachers and students of the cathedral schools*
There were two popes, one in Avignon and another in Rome.

50   *linear distances in equal periods of time*
p. 165.

50   *the times it takes to travel those distances*
If the object travels a distance s_1 in t_1 seconds and s_2 in t_2 seconds, then $s_1/s_2 = t_1/t_2$.

52   *worked on an idea that changed the world*
      We know very little about these people; even their names came by
      speculation. Heytesbury may have been the same person as Hogh-
      telbury or Heightilbury. And Richard Swineshead was confused
      with John Swineshead and sometimes with Roger Swineshead.
      This mathematician at Merton was also called "The Calculator" and
      also went by the name of Suiseth. He is reported to be the first person
      to show that an infinite series $1/2 + 2/4 + 3/8 + \ldots + n/2^n + \ldots$
      converges to a finite number. However, he didn't seem to use this
      in application to Zeno's dichotomy paradoxes.

52   *what was about to happen at Merton*
      Bradwardine's *Tractatus de Proportionibus Velocitatum* was written
      in 1328.

53   *delivering one of his lectures on motion*
      His book on motion is *Regule Solvendi Sophismata* (1335).

53   *are assumed to accelerate uniformly*
      Also called the *mean speed theorem*. Nicole Oresme proved this
      theorem sometime during the 1350s at the College of Navarre.
      For a geometric proof of the acceleration theorem, see *A Source
      Book,* pp. 243-53.

55   *velocity multiplied by the time of travel*

      The distance will be $\left(\dfrac{v + v_0}{2}\right)$, where v is the final velocity, v_0 is the

      initial velocity, and t is time. This is the same result we would get
      by using modern techniques.

58   *the chandelier would depend only on length*
      In 1656, the Dutch scientist, Christiaan Huygens, proved that the
      period of oscillation of a pendulum is equal to $2\neq \wp L/g$, where L is
      the length of the pendulum and g is the acceleration due to gravity.
      Therefore, the time of oscillation depends only on length.

59   *the rate and variation of the pulse*
      Sister Maria Celeste, *The Private Life of Galileo Compiled Princi-
      pally from His Correspondence and That of His Eldest Daughter*, ed-
      ited by Eugo Albéri and Carlo Aruini. Boston: Nichols & Noyes,
      1870, p. 17.

60   *the Caribbean to the port of Cádiz*
      Cochineal was thought to be part of a plant, but later discovered to
      be an insect inhabitant of a certain species of cactus. The insect

feeds from the plant in order to manufacture a pigment and store it in body fluids.

61   *and even the learned Padua*
Celeste, *The Private Life of Galileo,* pp. 17-8.

61   *set above more substantial attainments*
R. R. Palmer, *A History of the Modern World,* 2nd Ed. New York: Alfred Knopf, 1961, p. 54.

62   *mathematics taught me this method*
Galileo Galilei, *On Motion and On Mechanics,* translated by I. E. Drabkin and Stillman Drake. Madison: University of Wisconsin Press, 1960, p. 50. (Quotes on the next page are from this source.)

64   *the proportion to the rareness of the media*
Rareness here means the inverse of density. Air is *rarer* than water.

64   *the rareness of water is 4 and that of air is 16*
These numbers are just vague representations of the inverse of density. They are very far off from the actual numbers on any modern scale, but these numbers that Galileo gives serve the purpose. The actual rareness of water is $0.001 \text{m}^3/\text{kg}$; the rareness of air is $0.83 \text{m}^3/\text{kg}$.

64   *Suppose its velocity in air is 8*
This number also is made up. Note that Galileo is refuting Aristotle, so he can assume a constant velocity. Also note: The units for this velocity are arbitrary, since the end result will be unitless anyway.

65   *some medium such that its speed is 1*
Actually, this may not be true, but all Galileo needs is a medium whose rareness is between that of air and water. Benzene or ethanol would do.

65   *the ratios of velocities in two different media*
In modern notation, the ratios of velocities may be computed as $v_{1a}/v_{2b} = (w_1 - w_a)/(w_2 - w_b)$, where v is velocity, w weight, the subscript numbers indicate objects, and the subscript letters indicate medias a and b.

63   *with the agent that gave them motion?*
This is a paraphrasing of Aristotle's *Physics*, VIII. X. 266b 30.

64   *prime mover has ceased to move them*
Ibid.

66   *tried to take refuge in this view*
Galileo, *On Motion and on Mechanics,* p. 76.

67   *for there is nothing to move it*
Galileo is ignoring friction in the mechanism that permits it to rotate on its axis. He probably did not know about surface friction.

69   *assumed to be almost frictionless*
The frictionless requirement is of almost no consequence to the puzzle, yet it helps to imagine the chain as purely under the influence of gravity.

70   *even in the absence of friction*
Another way to see this is to watch how the 3, 4, 5 swaps proportions with the effective weights. We see this in the illustration below. The square block resting on the sloping surface represents the weight of that portion of the necklace that rests on the sloping surface. The smaller square block on the vertical side represents the weight of that portion of the necklace that hangs freely on the vertical side. Notice that the effective weight of the heavier block is exactly equal to the weight of the vertical block.

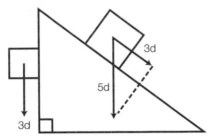

71   *perfect geometrical figures*
The World of Mathematics, vol. II, edited by James R. Newman. New York: Simon & Schuster, 1956, p. 728.

74   *suddenly acquires great speed*
Galileo Galilei, *Dialogues Concerning Two New Sciences*, translated by Henry Crew and Alfonso De Salvio. New York: The Macmillan Company, 1914, p. 161.

74–75 *squares of the times*
Ibid., p. 163.

75   *employed in traversing these distances*
Ibid., p. 174.

79   *epicycles were added*
The epicycle idea dates back the third century BCE, when Apollonius introduced them in his model. An epicycle may be thought of

as a circular orbit that itself orbits. The planet's orbit is a circle, but every once in a while it follows a smaller circular orbit around a point on the orbit. These epicycles were necessary to explain why we observe planets occasionally moving backward for short times. The appearance of backward movement is called retrograde motion.

83   *One now reads with amazement*
      The opening lines to Galileo's *Dialogue on the Great World Systems.*

85   *the image of the entire universe*
      See Plato's *Timaeus*, 54a-55c.

85   *only five such regular solids*
      They are the tetrahedron (with four sides), the cube (with six sides), the octahedron (with eight sides), the dodecahedron (with twelve sides), and the icosahedron (with twenty sides).

87   *my joy was to vanish into air*
      From *World of Mathematics,* p. 223.

87   *in dynamic laws*
      Hermann Weyl, *Symmetry*, Rev. ed. Princeton, New Jersey: Princeton University Press, 1983.

88   *prisons of the tables*
      Ibid., p. 228.

98   *has moved an equal distance*
      Ignoring Lorentz contraction.

104   *Hooke's Law*
      Hooke's law may be thought of in this way: If a one-pound weight stretches the spring x inches, then a two-pound weight will stretch it $2x$ inches. However, there are two caveats: 1) the law does not extend to stretching the spring beyond its restoring capacity, and 2) in practice, one inevitably finds imperfections in springs that cause small deviations from the ideal.

105   *in the eleventh century*
      Reputed to have been built in the year 1094. For an illustration and detailed description of how this clock worked, see David S. Landes, *Revolution in Time: Clocks and the Making of the Modern World.* Cambridge, Massachusetts: Harvard University Press, 1983, fig. 3 (after page 236).

105   *carrying out the said work*
      G.H. Baille, C. Clutton, C.A. Ilbert, *Britten's Old Clocks and Watches and Their Makers,* 7th Ed. New York: Bonanza Books, 1956. pp. 5-6.

106   *the verge escapement*
      Many different kinds of escapements have been invented and used

in clockmaking over the centuries. For an excellent account, see A.L. Rawlings, *The Science of Clocks and Watches*, edited by Timothy and Amyra Treffry. Upton, England: The British Horological Institute, 1944.

107   *the pendulum to make a complete swing*
The period of oscillation of a pendulum is approximately equal to $2\pi\sqrt{\dfrac{l}{g}}$, where l is the length of the pendulum and g is the acceleration due to gravity. The actual period is an infinite series given by

$$T = 2\pi\sqrt{\frac{l}{g}}\left\{1 + \left(\frac{1}{2}\right)^2 \sin^2 \frac{1}{2}\alpha + \left(\frac{1.3}{2.4}\right)^2 \sin^4 \frac{1}{2}\alpha + \ldots\right\}, \text{ where}$$

the ellipsis at the right end means that there are an infinite number of terms in ascending order of even powers of $\sin\dfrac{1}{2}\alpha$. For small swings, the powers of $\sin\dfrac{1}{2}\alpha$ are very small, so all terms beyond the first are insignificant.

107   *grows with amplitude*
An alternative would be to fix a weight to the top end of a flexible steel band. Then fix the bottom end so the weight can sway back and forth. This may look like an upside-down pendulum, but the path of the weight would not be a circular arc. It would be an *isochronous* curve.

109   *unit of time that cannot be split*
The history of sports has seen astounding examples of hairsplitting times. Two cross-country skiers, who had been racing for hours, finished one-hundredth of a second apart at the 1990 Lake Placid Winter Olympics.

114   *planes that cut cones*
The circle was thought of as a collection of points determined by a specific rule. Euclid defined the circle as "a plane figure contained by one line such that all the straight lines falling upon it from one point [the center] among those lying within the figure are equal to one another."

114   *without the permission of the other*
For example, the coordinates of a point (x,y) on a parabola pass-

ing through a specified point (a,b) are related by the equation $y = ax^2 + b$.

116 *there corresponds a unique value of y*
The variable x should be restricted to some interval.

117 *Achilles overtakes the tortoise*
Opus Geometricum Quadraturae Circuli et Sectionum Coni (1647).

117 *infinite division of the number line*
From Gregory's work on conics, Leibniz, Fermat, and Descartes later credited him as one of the founders of analytic geometry.

118 *but dark and penetrating eyes*
In the following, quoting John Theodore Merz, a Leibniz biographer: Hal Hellman, *Great Feuds in Science, Ten of the Liveliest Disputes Ever*. New York: John Wiley & Sons, 1998, p. 41.

119 *romantic flight in check*
Tobias Dantzig, *Number: The Language of Science*, edited by Joseph Mazur. New York: Pi Press, 2005, p. 135.

125 *attempt to square the circle*
See George Johnston Allman, *Greek Geometry from Thales to Euclid*. Dublin: Hodges, Figgis, & Co., 1889, p. 66: "... he formed a polygon of twice as many sides; and doing the same again and again, until he had exhausted the surface, he concluded that in this manner a polygon would be inscribed in the circle, the sides of which, on account of their minuteness, would coincide with the circumference of the circle."

127 *figure with finite area*
Fermat, Oresme, and Roberval also exhibited such figures.

127 *depends on the size of the interval*
Galileo had reasoned this same way in his *Two New World Systems,* in which he showed how to analyze accelerated motion by thinking of it as synthesized from infinitesimally small cases of uniform motion. I thank Emily Grosholz for informing me that Galileo borrowed this idea and improved it from the Oxford kinematicists.

133 *failure to appreciate this fact*
Carl Boyer, *The History of the Calculus and Its Conceptual Development*. New York: Dover, 1949, p. 295.

133 *infinite geometric series*
An *infinite geometric series* is an infinite sum of terms in which each successive term of the sum decreases in proportion to a power of the

term before it. The series $a + ar^2 + ar^3 + \ldots = \dfrac{a}{1-r}$ as long as

$-1 < r < 1$.

For example, $1 + \left(\dfrac{1}{2}\right)^2 + \left(\dfrac{1}{2}\right)^3 + \ldots = \dfrac{1}{1-\left(\dfrac{1}{2}\right)} = 2$.

133   *place when it will happen*
      Florian Cajori, *A History of Mathematics*. New York: Chelsea, 1985, pp. 181-2.

137   *thither they return again*
      Ecclesiastes 1:5-7.

138–139 *Cycle and epicycle, orb in orb . . .*
      John Milton, *The Portable Milton*, edited by Douglas Bush. New York: Viking, 1961, pp. 416-7.

139   *body has in virtue of its mass*
      By "inertial motion" I mean the motion a body has by virtue of its mass.

140   *forces exerted on the body*
      Because force equals mass times acceleration.

140   *as its mathematical agent*
      Remember, this law says that the force exerted between two bodies is inversely proportional to the square of the distance between them (the masses being the constant of proportionality).

142   *carries a planet in its orbit*
      From *World of Mathematics*, p. 140.

143   *it goes roughly as follows*
      The dialogue of the story is roughly taken from Herbert Warren Turnbull's essay, *The Great Mathematicians*, appearing in *World of Mathematics*, p. 144.

144–145 *the earth draws the apple*
      From William Stukeley's biography of Sir Isaac Newton, Royal Society manuscript #142, transcribed from original by Rob Iliffe, September 2004, as part of the Newton Project of Imperial College, London. URL: http://www.newtonproject.ic.ac.uk/texts/rsstukeley_n.html.

146   *to be made more publick*

Sir Isaac Newton, *The Mathematical Papers of Isaac Newton,* Vol. VII, edited by W.T. Whiteside. Cambridge: Cambridge University Press, 1976, pp. 8-9n.

146 *John Conduitt and William Stukeley*

These stories are well documented by John Conduitt and William Stukeley in Royal Society manuscripts, which are readily available through the Newton Project of Imperial College, London. URL.: http://www.newtonproject.ic.ac.uk/viewcp.html.

146–147 *more than at any time since*

From E.N. Da C. Andrade's article "Isaac Newton" in *World of Mathematics,* p. 257.

151 *the latest scientific discoveries*

For a wonderful history and description of early coffeehouses, see Tom Standage, *History of the World in 6 Glasses.* New York: Walker, 2005, pp. 141-172.

151 *volumes of the first encyclopedia*

It took twenty-one years to complete.

155 *strange seas of thought, alone*

William Wordsworth, *The Prelude, Selected Poems and Sonnets,* edited by Wayne Booth. New York: Holt, Rinehart and Winston, 1954, p. 238.

157 *light from droplets of rain*

Raindrops are involved, but the cause is refraction, not reflection.

157 *the speed of light fast but finite*

The mathematician Abu Ali al-Hasan Ibn al-Haitham, otherwise known as al Haythen (965–1040), believed the speed of light to be finite, but had no way of experimentally proving it.

157 *Sagredo, Simplicio, and Salviati debate*

Sagredo is the neutral figure posing the questions; Simplicio is an Aristotelian; and Salviati is the figure representing Galileo's opinions.

158 *instantly see that of the other*

Galileo, *Dialogues Concerning Two New Sciences,* p. 43.

159 *light had to travel to Earth*

There is a discrepancy between Roemer's observation of a twenty-two-minute lag time and what the lag time should have been using today's knowledge of the speed of light and the earth's closest and furthest distance from Jupiter. Using the established speed of 300,000 km/sec, and knowing that the difference between the earth's closest and furthest distance from Jupiter is 300,000,000 km, the lag time would be a bit more than 16½ minutes.

159   *Huygens*
Christiaan Huygens was the Dutch scientist who built the first pendulum clock in 1656. Its constancy of motion increased the accuracy of time tenfold.

164   *reinforcing or canceling one another*
See Thomas Young, "Experimental Demonstration of the General Law of the Interference of Light," *Philosophical Transactions of the Royal Society of London*, vol. 94 (1804).

166   *invisible lines of force*
For an excellent readable account, see Robyn Arianrhod, *Einstein's Heroes: Imagining the World Through the Language of Mathematics*. New York: Oxford, 2005.

175   *speed is the same*
Today we know that the speed of light in a vacuum is more accurately 299,792,458 m/sec.

175   *floating in water along with the ether*
Note that we are using the word *moving* in a peculiar way: By its definition, the ether is stationary, so the movement of anything such as the earth is movement relative to the ether.

177   *distance divided by velocity*
The time it takes to make the trip is $\dfrac{d}{c - v}$. The return time is

$\dfrac{d}{c + v}$. So the total round trip time is $\dfrac{d}{c - v} + \dfrac{d}{c + v}$, which, with a bit of algebra, simplifies to $\dfrac{2dc}{c^2 - v^2}$.

179   *less than ten inches long*
It is important to notice that we are talking about objects and light traveling through a vacuum. Objects in air can travel faster than the speed of light in air, but not faster than the speed of light in a vacuum. Nothing can travel faster than the speed of light in a vacuum.

179–180   *object would have a negative length*
Actually, for speeds greater than the speed of light, the FitzGerald formula would give the length as an imaginary number.

180   *the percentage of FitzGerald's shrinking*
For electrons, the Lorentz mass increase can be measured by measuring the ratio of mass to charge. The mass/charge ratio can be mea-

sured by its deflection in a magnetic field. Any increase in this ratio due to an increase in velocity must be due to an increase in mass, since the charge doesn't increase. It turns out that the mass/charge ratio does increase precisely by what Lorentz's equation expects.

180   *ratio did increase with velocity*
In 1899, the British physicist Joseph John Thomson (better known as J.J. Thomson) was able to measure the electric charge itself and thereby measure the mass of charged particles directly. These experiments with electrons could not have been made without an ingenious method for creating a vacuum in a glass tube perfected by Heinrich Geissler, a glassblower and experimental physicist. For a more detailed account of this, see John Gribbin, *The Scientists: A History of Science Told Through the Lives of Its Greatest Inventors.* New York: Random House, 2002, pp. 490-3.

186   *christen the world*
Hendrik A. Lorentz, Albert Einstein, and Hermann Weyl, *The Principle of Relativity*. New York: Dover, 1952.

186   *a relative frame of reference*
Normally when a variable is dependent on others it does not increase the dimension. However, in this case, the time variable depends on the frame of reference. This complicates things globally, but locally the time variable contributes an extra dimension.

189   *his teacher Hermann Minkowski*
Gribben, *The Scientists,* p. 393.

190   *—is relative*
This paper is translated into English in *Einstein: A Centenary Volume*, edited by A.P. French. Cambridge, Massachusetts: Harvard University Press, 1979, p. 281.

190   *previously never imagined*
Paul Davies, *About Time: Einstein's Unfinished Revolution*. New York: Touchstone, 1996, p. 47.

190   *with reference to our sensations*
Letter from Albert Einstein to Herbert Samuel, Oct. 13, 1950. Printed in Herbert L. Samuel, *Essay in Physics*. New York: Harcourt, Brace, 1950, p. 158.

190   *—it has tangents*
The term smooth is a mathematical term that roughly means that all higher-order derivatives exist. For the purposes of this book, we mean that it has no sharp edges, no punctures, no spikes, and no severed edges.

191   *the affairs of sound business men*
H.G. Wells, *The Outline of History*, Vol. II. Garden City, New York: Garden City Books, 1920, p. 829.

194   *precisely where the object will be*
There are some problems when it comes to more complicated motion such as a system of bodies under gravitation. For example, the famous generalized three-body problem does not have an exact closed-form solution. But the kinds of motion problems that Zeno introduced have trivial prediction functions enabling us to know the exact place an object will be at when the time is specified.

195   *particular position in space and time*
Actually, the wave function Ψ itself cannot give a probability. It is just a sum of sine functions. However, the square of a wave function can. $\Psi^2(x, y, z, t)$ measures the probability of finding a particle at position (x, y, z) at time t.

196   *wavelength of the radiation*
The German theoretical physicist Wilhelm Wien showed this in 1893 for the ideal situation of *black body* radiation, where light is absorbed or radiated perfectly.

197   *it is as discontinuous as water*
To give an impression of the size, $h = 6.63 \times 10^{-34}$ joule-seconds, and 1 joule is equal to the amount of energy needed to lift 0.738 pounds 1 foot.

198   *suppose it continuous instead*
Louis De Broglie, *Matter and Light: The New Physics*, translated by W.H. Johnston. New York: Dover, 1939, p. 220.

199   *uses that energy to escape*
Albert Einstein, "Uber einen die Erzeugung und Umwandlung des Lichtes betreffenden heuristichen Standpunkt (On a Heuristic Viewpoint Concerning the Generation and Transformation of Light)," *Annalen der Physik,* 17, 1905, pp. 132-84.

199   *will not make a difference*
The term *photon*—which was not used before 1926—is now used in place of *light quantum*.

200   *generated only as complete units*
As quoted in Gribbin's wonderful book, *The Scientists*, p. 511.

200   *Planck's constant h*
The *momentum* of a moving object is defined as the mass times velocity. It is related to the force imposed on the object in the sense that the force is equal to the rate of change in momentum.

The *wavelength* of a wave is the distance in which the wave repeats itself.

203   *from fundamental particles*

The four known fundamental forces of nature are:

Gravity—by far the weakest force.

Electromagnetic—holds atoms, molecules, solids, and liquids together.

Weak nuclear —a very short-range force that permits neutron and proton decay and the fusion process that occurs in stars.

Strong nuclear—the strongest of all forces, responsible for holding the nucleus together.

204   *to understand string theory*

Brian Greene, *The Elegant Universe.* New York: Norton, 1999, p. 139.

208   *roots of polynomial equations*

Kummer surfaces

208   *to measure within the space*

For the mathematically inclined: In differential geometry, *Ricci-flat* means that the trace of the Riemann curvature tensor (the multidimensional matrix that classifies how sharply curved the space is at every point independent of the reference frame) is always zero.

209   I am indebted to Professor Andrew J. Hanson of Indiana University for the image of this two-dimensional cross section of the six-dimensional Calabi-Yau manifold. For more information see Professor Hanson's home page: http://www.cs.indiana.edu/~hanson.

212–213   *ceases forthwith to be itself*

All quotes on these pages are from William James, *The Principles of Psychology*, Vol. 1. New York: Dover, 1950, p. 244.

213   *long enough for the next to take over*

See Hermann von Helmholtz, *Handbook of Physiological Optics,* translated by J.P.C. Southhall. Leipzig: Leopold Voss, 1925, p 372. Also see Chapter 6, "Recent Progress of the Theory of Vision," in Hermann von Helmoltz, *Science and Culture: Popular and Philosophical Essays*, edited by David Cahan. Chicago: University of Chicago Press, 1995, p. 127.

218   *unless acted upon and F = ma*

Here the *a* represents the acceleration due to gravity.

218–219   *no rational explanation for it*

E. Wigner, "The Unreasonable Effectiveness of Mathematics in the Natural Sciences," *Communication in Pure and Applied Mathematics*, vol. 13, no. I (February 1960).

Further Reading

Allman, George Johnston, *Greek Geometry from Thales to Euclid*. New York: Arno Press, 1976.

Aquinas, St. Thomas, *Commentary on Aristotle's Physics*, translated by Richard J. Blackwell, Richard J. Spath, and W. Edmund Thirlkel. London: Routledge & Kegan Paul, 1963.

Arianrhod, Robyn, *Einstein's Heroes: Imagining the World Through The Language of Mathematics*. New York: Oxford, 2005.

Aristotle, *The Physics*, 2 vols., translated by Philip H. Wicksteed and Francis M. Cornford. Cambridge, Massachusetts: Harvard University Press, 1934.

Aristotle, *On the Heavens*, translated by W.K.C. Guthrie. Cambridge, Massachusetts: Harvard University Press, 1939.

Baille, G.H., C. Clutton, and C.A. Ilbert, *Britten's Old Clocks and Watches and Their Makers*, 7th Ed. New York: Bonanza Books, 1956.

Barnes, Jonathan, trans. and ed., *Early Greek Philosophy*. London: Penguin, 1987.

Bietkowski, Henryk, and Wlodzimierz Zonn, *The World of Copernicus*, translated by Doreen Heaton-Potworowska. Warsaw: Arkady, 1973.

Black, Max, *The Nature of Mathematics: A Critical Survey*. London: Routledge & Kegan Paul, 1933.

Boardman, John, and Jasper Griffin, eds., *The Oxford History of the Classical World*. New York: Oswyn Murray, Oxford University Press, 1986.

Bochner, Salomon, *The Role of Mathematics in the Rise of Science*. Princeton, New Jersey: Princeton University Press, 1966.

Bohr, Niels, "On the Application of the Quantum Theory to Atomic Structure," *Proceedings of the Cambridge Philosophical Society*, reprinted in translation by F.L. Curtiss. Cambridge: Cambridge University Press, 1924.

———— "Can Quantum-Mechanical Description of Physical Reality be Considered Complete?" *Physical Review* vol. 48 (1935): 696–702.

Bolles, Edmund Blair, ed., *Galileo's Commandment, 2,500 Years of Great Science Writing*. New York: W.H. Freeman, 1999.

Bolzano, Bernard, *Paradoxes of the Infinite*, translated by Dr. Fr. Prihonsky. London: Routledge & Kegan Paul, 1950.

Boorstin, Daniel J., *The Discoverers*. New York: Random House, 1983.

Boyer, Carl B., *The History of the Calculus and Its Conceptual Development*. New York: Dover, 1959.

Bridgman, P.W., *The Logic of Modern Physics*. New York: The Macmillan Company, 1961.

————, *A Sophisticate's Primer of Relativity*. Middletown, Connecticut: Wesleyan University Press, 1962.

Buchanan, Scott, *Truth in the Sciences*. Charlottesville, Virginia: University Press of Virginia, 1972.

Burke, Kenneth, *Attitudes Toward History*. Boston: Beacon Press, 1961.

Bynum, W.F., E.J. Browne, and Roy Porter, eds., *Dictonary of the History of Science*. Princeton, New Jersey: Princeton University Press, 1981.

Cahan, David, ed., *Science and Culture: Popular and Philosophical Essays*. Chicago: University of Chicago Press, 1995.

Cajori, Florian, "The History of Zeno's Arguments on Motion: Phases in the Development of the Theory of Limits," *The American Mathematical Monthly*, vol. 22, no. 1 (January 1915).

————, *A History of Mathematics*. New York: Chelsea, 1985.

Cardwell, Donald, *The Norton History of Technology*. New York: Norton, 1994.

Celeste, Sister Maria, *The Private Life of Galileo Compiled Principally from His Correspondence and That of His Eldest Daughter*, edited by Eugo Albéri and Carlo Aruini. Boston: Nichols and Noyes, 1870.

Clagett, Marshall, *Greek Science in Antiquity*. Plainview, New York: Books for Libraries Press, 1955.

Cohen, Morris R., and I.E. Drabkin, *A Source Book in Greek Science*. Cambridge, Massachusetts: Harvard University Press, 1948.

Crawley, Richard, trans., *The Complete Writings of Thucydides: The Peloponnesian War*. New York: Modern Library, 1951.

Dantzig, Tobias, *Number: The Language of Science*, 4th Ed. Garden City, New York: Doubleday, 1956.

Dauben, Joseph, *Georg Cantor: His Mathematics and Philosophy of the Infinite*. Princeton, New Jersey: Princeton University Press, 1990.

Davies, Paul, *About Time: Einstein's Unfinished Revolution*. New York: Touchstone, 1996.

De Broglie, Louis, *Matter and Light: The New Physics*, translated by W.H. Johnston. New York: Dover, 1939.

D'Espagnat, Bernard, "The Quantum Theory and Relativity," *Scientific American*, November 1979.

Dehaene, Stanislas, *The Number Sense: How the Mind Creates Mathematics*. New York: Oxford University Press, 1997.

De Santillana, Giorgio, *The Crime of Galileo*. Chicago: The University of Chicago Press, 1955.

Dijksterhuis, E. J., *Archimedes*. Princeton, New Jersey: Princeton University Press, 1987.

Dodds, E.R., *The Greeks and The Irrational*. Berkeley, California: University of California Press, 1959.

Drabkin, I.E., and Stillman Drake, eds., *Galileo Galilei on Motion and On Mechanics*. Madison, Wisconsin: The University of Wisconsin Press, 1960.

Durell, Clement, *Readable Relativity: A Book For Non-Specialists*. New York: Dover, 2003.

Eddington, A.S., *The Mathematical Theory of Relativity*. New York: Chelsea, 1975.

Einstein, Albert, *Relativity*. New York: Pi Press, 2005.

Farrington, Benjamin, *Greek Science: Its Meaning for Us*. Harmondsworth, England: Penguin, 1966.

Feeney, D.C., *The Gods in Epic: Poets and Critics of the Classical Tradition*. Oxford: Clarendon Press, 1991.

Fölsing, Albrecht, *Albert Einstein, A Biography*, translated by Ewald Osers. New York: Viking, 1997.

Fowler, David, *The Mathematics of Plato's Academy*. London: Oxford University Press, 1987.

French, A., *The Growth of the Athenian Economy*. London: Routledge & Kegan Paul, 1964.

Galilei, Galileo, *Dialogues Concerning Two New Sciences*, translated by

Henry Crew and Alfonso de Salvio. New York: The Macmillan Company, 1914.

Galison, Peter, *Einstein's Clocks, Poincaré's Maps: Empires of Time.* New York: Norton, 2004.

Garner, Jane F., and Thomas Wiedemann, *The Roman Household: A Sourcebook.* London, Routledge, 1991.

Geymonat, Ludovico, *Galileo Galilei: A Biography and Inquiry into His Philosophy of Science*, translated by Stillman Drake. New York: McGraw Hill, 1965.

Goodstein, R. L., *Essays in the Philosophy of Mathematics.* Leicester: Leicester University Press, 1965.

Gott, J. Richard, *Time Travel in Einstein's Universe: The Physical Possibilities of Travel Through Time.* Boston: Mariner, 2002.

Grant, Edward, ed., *A Source Book for Medieval Science*, Cambridge, Massachusetts: Harvard University Press, 1974.

Graves, Robert, *The Greek Myths Complete Edition.* London: Penguin, 1960.

Greenberg, Noah, W.H. Auden, and Chester Kallman, eds., *An Elizabethan Song Book.* Garden City, New Jersey: Doubleday, 1955.

Greenblatt, Steven, *Will in the World, How Shakespeare Became Shakespeare.* New York: Norton, 2004.

Gribbin, John, *The Scientists: A History of Science Told Through The Lives of Its Greatest Inventors.* New York: Random House, 2002.

Grünbaum, A., *Modern Science and Zeno's Paradoxes.* Middletown, Connecticut: Wesleyan University Press, 1967.

Hall, Alfred Rupert, *Philosophers at War: The Quarrel between Newton and Leibniz.* Cambridge: Cambridge University Press, 2002.

Hamilton, Edith, and Huntington Cairnes, eds., *The Collected Dialogues of Plato*, translated by Francis Macdonald Cornford. Princeton, New Jersey: Princeton University Press, 1961.

Heath, Sir Thomas, *A History of Greek Mathematics*, 2 vols. Oxford: Clarendon Press, 1921.

———, *Aristarchus of Samos: The Ancient Copernicus.* New York: Dover, 1981.

Hellman, Hal, *Great Feuds in Science, Ten of the Liveliest Disputes Ever.* New York: John Wiley & Sons, 1998.

Hersh, Davis, *Descartes' Dream.* Orlando, Florida: Harcourt Brace Jovanovich, 1986.

Hesiod, *Theogeny and Works and Days*, translated by M. L. West. Oxford: Oxford University Press, 1988.

Holt, Jim, "Time Bandits," *The New Yorker,* February 28, 2005.

Holton, Gerald, *Introduction to Concepts and Theories in Physical Science*. Reading, Massachusetts: Addison-Wesley, 1952.

————, *The Scientific Imagination: Case Studies*. Cambridge: Cambridge University Press, 1978.

Hoyle, Fred, *Nicolaus Copernicus*. New York: Harper & Row, 1973.

Huggett, N., ed., *Space from Zeno to Einstein: Classic Readings with a Contemporary Commentary*. Cambridge, Massachusetts: MIT Press, 1999.

Hume, David, *An Inquiry Concerning Human Understanding*, edited by Charles W. Hendel. New York: Macmillan, 1955.

Jourdain, Philip E.B., "The Flying Arrow: An Anachronism," *Mind*, New Series, vol. 25, no. 97 (January 1916).

Kelman, P.J., and Spelke, E.S., "Perception of Partly Occluded Objects in Infancy." *Cognitive Psychology*, vol. 15 (1983) 483–524.

Kirk, G.S., and J.E. Raven, *The Presocratic Philosophers: A Critical History with a Selection of Texts*. New York: Cambridge University Press, 1962.

Klein, Arthur H., "Pieter Bruegel the Elder as a Guide to 16th-Century Technology," *Scientific American*, vol. 238, no. 3 (March 1978).

Kleene, Stephen Cole, *Introduction to Mathematics*. Princeton, New Jersey: Van Nostrand, 1962.

Laertius, Diogenes, *Lives of Eminent Philosophers*, 2 vols, translated by R.D. Hicks. Cambridge, Massachusetts: Harvard University Press, 1925.

Landes, David S., *Revolution in Time: Clocks and the Making of the Modern World*. Cambridge, Massachusetts: The Belknap Press of Harvard University Press, 1983.

Lattimore, Richard, trans., *The Odyssey of Homer*. New York: Harper & Row, 1965.

————, *The Iliad of Homer*. Chicago: University of Chicago Press, 1951.

Lee, H. D. P., ed., *Zeno of Elea*. Amsterdam: Adolf Hakkert, 1967.

Lightman, Alan P., *Time Travel and Papa Joe's Pipe*. New York: Charles Scribner's Sons, 1984.

Lloyd, G.E.R., *Early Greek Science: Thales to Aristotle*. New York: Norton, 1970.

Lucretius, *The Nature of the Universe*, translated by R.E. Latham. Baltimore, Maryland: Penguin, 1951.

Matson, W. I., "Zeno Moves!," in *Essays in Ancient Greek Philosophy*. Vol. 6, *Before Plato*, edited by A. Preus. Albany: SUNY Press, 2001.

Mach, Ernst, *The Analysis of Sensations and the Relation of the Physical to the Psychical*, translated by C.M Williams. New York: Dover, 1959.

————, *Knowledge and Error*, translated by Thomas J. McCormack. Boston: D. Reidel, 1976.

Marchall, Clagett, *Science of Mechanics in the Middle Ages*. Madison, Wisconsin: University of Wisconsin Press, 1959.

Mason, Stephen F., *A History of the Sciences*. New York: Collier, 1962.

Maxwell, J.C., "A Dynamical Theory of the Electromagnetic Field," *Philosophical Transactions*, vol. 166 (1865): 495–512; reprinted in *The Scientific Papers of James Clerk Maxwell*, Vol. 1. New York: Dover, 1952.

Maxwell, J.C., *Nature,* vol. 2 (September 22, 1870).

Mazur, Joseph *Euclid in the Rainforest: Discovering Universal Truth in Logic and Math*. New York: Pi Press, 2005.

McLaughlin, W. I., "Resolving Zeno's Paradoxes," *Scientific American*, November 1994, pp. 84–89.

Michelson, Dorothy Livingston, *The Master of Light: A Biography of Albert A. Michelson*. New York: Charles Scribner's Sons, 1973.

Milton, John, *The Portable Milton*, edited by Douglas Bush. New York: Viking, 1961.

Munro, Dana C., "The Speech of Pope Urban II at Clermont, 1095," *American Historical Review*, vol. 11, no. 1 (October 1905).

Murray, Gilbert, *The Rise of the Greek Epic*. Oxford: Clarendon Press, 1907.

————, *Five Stages of Greek Religion*. Garden City, New York: Doubleday Anchor, 1955.

Murray, Oswyn, *Early Greece*. Stanford, California: Stanford University Press, 1983.

Nagel, Ernest, *The Structure of Science: Problems in the Logic of Scientific Explanation*. New York: Harcourt Brace & World, 1961.

Neugebauer, Otto, *The Exact Sciences in Antiquity*. Providence, Rhode Island: Brown University Press, 1957.

Newman, James R., ed., *The World of Mathematics*, Vol. II. New York: Simon & Schuster, 1956.

Norsen, Travis, "Einstein's Boxes," *American Journal of Physics*, vol. 73, no. 2 (February 2005).

Oerter, Robert, *The Theory of Almost Everything: The Standard Model, the Unsung Triumph of Modern Physics*. New York: Pi Press, 2006.

Overbye, Dennis, "Quantum Trickery: Testing Einstein's Strangest Theory," *New York Times*, December 27, 2005.

Palmer, R.R., *A History of the Modern World*, 2nd Ed. New York: Alfred A. Knopf, 1961.

Plutarch, *The Rise and Fall of Athens: Nine Greek Lives*, translated by Ian Scott-Kilvert. Baltimore, Maryland: Penguin, 1960.

———, *The Lives of the Noble Grecians and Romans*, translated by John Dryden. New York: The Modern Library, 1932.

Poincaré, Henri, *Science and Hypothesis*, translated by W.J.G. Mineola, New York: Dover, 1952.

———, *The Value of Science*, translated by George Bruce Halsted. New York: Dover, 1958.

———. *Science and Method*, translated by Francis Maitland. New York: Dover, 1952.

Pollio, Marcus Vitruvius, *The Ten Books on Architecture,* translated by Marris Hicky Morgan. New York: Dover, 1960.

Proclus, *A Commentary on the First Book of Euclid's Elements*, translated with introduction and notes by Glenn R. Morrow. Princeton, New Jersey: Princeton University Press, 1970.

Randel, Don Michael, ed., *The Harvard Dictionary of Music*. Cambridge, Massachusetts: The Belknap Press of Harvard University Press, 2003.

Rawlings, A.L., *The Science of Clocks and Watches*, edited by Timothy and Amyra Treffry. Upton, England: The British Horological Institute, 1994.

Rawlings, George, ed., *The History of Herodotus*, translated by E. H. Brakeney. New York: Dutton, 1910.

Reynolds, L.D., and N.G. Wilson, *Scribes and Scholars: A Guide to the Transmission of Greek and Latin Literature*. Oxford: Clarendon Press, 1968.

Ronan, Colin, *Galileo*. New York: G.P. Putnam's Sons, 1974.

Rucker, Rudy, *Infinity and the Mind*. Princeton, New Jersey: Princeton University Press, 1995.

Russell, Bertrand, *Our Knowledge of the External World As a Field for Scientific Method in Philosophy*. Chicago: Open Court, 1914.

———, *Introduction to Mathematical Philosophy*. New York: Dover, 1993.

Sacks, Oliver, "The Mind's Eye, What the Blind See," *The New Yorker*, July 28, 2003.

———, "Speed: Aberrations of Time and Movement," *The New Yorker*, August 23, 2004.

Salmon, W. C., *Zeno's Paradoxes*. Indianapolis, Indiana: Hackett, 2001.

Samuel, Herbert L., *Essay in Physics*. New York: Harcourt Brace & Company, 1952.

Sanders, J.H., *Velocity of Light*. Oxford: Pergamon Press, 1965.

Sanderson, Edgar, J.P. Lamberton, and John McGovern, *Six Thousand*

Years of History. Vol. 4, *Great Philosophers*. Philadelphia: E. R. Du-Mont, 1900.

Shea, William R., and Mariano Aritgas, *Galileo in Rome: The Rise and Fall of a Troublesome Genius*. New York: Oxford University Press, 2003.

Smith, D.E., *History of Mathematics,* 2 vols. New York: Dover, 1958.

Standage, Tom, *A History of the World in 6 Glasses*. New York: Walker & Company, 2005.

Swenson, Loyd S. Jr., *The Ethereal Aether: A History of the Michelson-Morley Aether-Drift Experiments, 1880–1930*. Dissertation, The Claremont Graduate School, Claremont, California, 1962.

Tannery, P., "Le concept scientifique du continu: Zénon d'Elée et Georg Cantor," *Revue Philosophique de la France et de l'Etranger,* 20 (1885): 385.

Thomas, Ivor, trans., *Selections Illustrating the History of Greek Mathematics in Two Volumes*. Vol. 1, *From Thales to Euclid*. Cambridge, Massachusetts: Harvard University Press, 1957.

Thompson, Benjamin (Count Rumford), *Philosophical Transactions,* vol. 88 (1798).

Thomson, George, *Studies in Ancient Greek Society*. New York: The Citadel Press, 1965.

Thucydides, *The Peloponnesian War*, edited by A.P. Peabody, translated by B. Jowett. Boston: D. Lothrop & Co., 1883.

Tolstoy, Leo, *War and Peace*, translated by Constance Garnett. New York: Random House, 1940.

Untermeyer, Louis, ed., *This Singing World*. New York: Harcourt Brace & Company, 1923.

Vlastos, G., "Zeno of Elea," in *The Encyclopedia of Philosophy*, edited by P. Edwards. New York: The Macmillan Co. and The Free Press, 1967.

Von Helmholtz, Hermann, *Handbook of Physiological Optics*, translated by J.P.C. Southhall. Leipzig: Leopold Voss, 1925.

Wells, H.G., *The Outline of History*, 2 vols. Garden City, New York: Garden City Books, 1920.

———, *The Time Machine and The War of the Worlds*. Ann Arbor, Michigan: Borders, 2004.

Weyl, Hermann, *Symmetry*, Rev. Ed. Princeton, New Jersey: Princeton University Press, 1983.

White, Andrew D., *A History of the Warfare of Science with Theology in Christendom*. New York: George Braziller, 1955.

Whitehead, A. N., *Process and Reality*. New York: Macmillan, 1929.

———, *An Introduction to Mathematics*. New York: Henry Holt and Company, 1939.

Whiteside, W.T., ed., *The Mathematical Papers of Isaac Newton*. Cambridge: Cambridge University Press, 1976.

Wigner, Eugene, "The Unreasonable Effectiveness of Mathematics in the Natural Sciences," *Communication in Pure and Applied Mathematics*, vol. 13, no. 1 (February 1960).

Wilson, Curtis, "Kepler and the Mode of Vision," *The St. Johns Review*, vol. 32, no. 1 (July 1980).

Young, Thomas, "Experimental Demonstration of the General Law of the Interference of Light," *Philosophical Transactions of the Royal Society of London*, vol. 94 (1804).

Index

Note: Page numbers in *italics* refer to illustrations.